South African Air Force

The Flying Springboks 1939–80

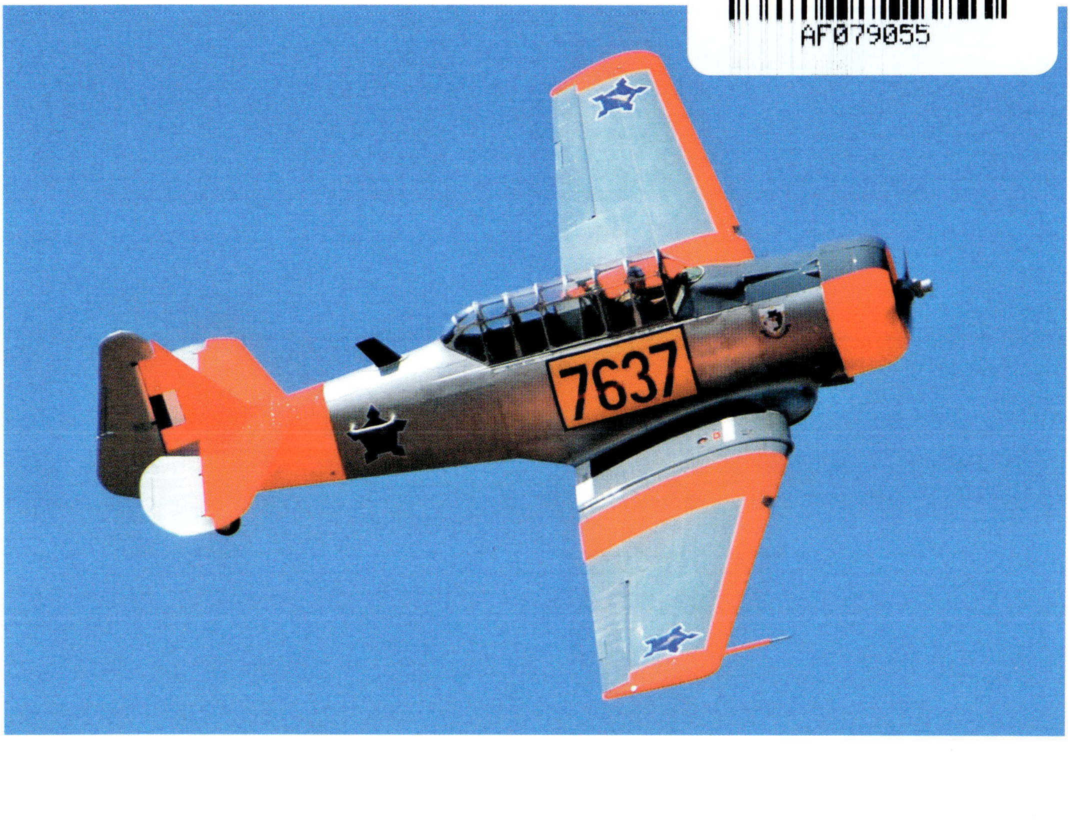

GERRY VAN TONDER

KEY Books

AIR FORCES SERIES, VOLUME 13

Front cover image: A seminal moment in the post-war history of the SAAF occurred in 1950 when delivery was taken of ten de Havilland Vampires, marking the transition of the air force's frontline fighter capabilities from propeller-driven aircraft to jet engines. Replacing the aging Spitfire Mk IXs at AFB Waterkloof, the iconic twin-boomed 'Vamp' served two decades before being phased out of operational duties by the end of 1972. (Ossewa, CC BY-SA 4.0, Wikimedia Commons)

Title page image: Former SAAF 7637, this North American Harvard (T-6 Texan) sports the emblem of SAAF No. 40 Squadron. (Ossewa, CC BY-SA 4.0, Wikimedia Commons)

Contents page image: Sud Aviation/Aérospatiale Puma. (SAVA)

Back cover image: Maritime patrol aircraft Avro Type 716 Shackleton MR3 1717, on static display at Air Force Base (AFB) Ysterplaat, Cape Town. (Oleg V. Belyakov, CC BY-SA 3.0, Wikimedia Commons)

Acknowledgements

The author drew inspiration from Willem Steenkamp and the late Herman Potgieter's magnificent *Aircraft of the South African Air Force* (C. Struik, Cape Town, 1980) and Hanlie Snyman Wroth and Gerry van Tonder's *North of the Red Line, Recollections of the Border War by Members of the SADF and SWATF 1966–1989* (30° South Publishers, Pinetown, 2016).

Grateful thanks to Colonel Dudley Wall for his exclusive World War Two SAAF aircraft drawings, used where period colour photographs of good quality were unavailable. Col Wall also generously provided images of his private collection of embroidered SAAF squadron patches featured in Chapter 3.

I extend my special thanks to Bruce Ross Strachan, Founding President of the South African Veteran Association (SAVA) for so magnanimously assisting with my sourcing of some of the spectacular images that appear in this book (https://sadf.info/ArmourmentsAirforce.html). Thanks to good friend Andy Jackson for his outstanding restoration work on some of the old photos.

Finally, to the SAAF and SADF veteran communities at large, thank you. Any omission of attribution is unintentional as many of the images I sourced online do not display any credits.

Published by Key Books
An imprint of Key Publishing Ltd
PO Box 100
Stamford
Lincs PE9 1XP

www.keypublishing.com

The right of Gerry van Tonder to be identified as the author of this book has been asserted in accordance with the Copyright, Designs and Patents Act 1988 Sections 77 and 78.

Copyright © Gerry van Tonder, 2024

ISBN 978 1 80282 956 3

All rights reserved. Reproduction in whole or in part in any form whatsoever or by any means is strictly prohibited without the prior permission of the Publisher.

Typeset by SJmagic DESIGN SERVICES, India.

Contents

Introduction 4

Chapter 1 History 6

Chapter 2 Aircraft 12
1. de Havilland Tiger Moth ... 18
2. Hawker Fury 19
3. Bücker Jungmann 20
4. Hawker Hart 21
5. Junkers Ju 52/3m 22
6. Junkers Ju 86Z 23
7. Avro Tutor 24
8. Hawker Audax 25
9. Hawker Hartebees 26
10. Hawker Hind 27
11. Westland Wapiti 28
12. Airspeed Oxford 29
13. Avro Anson 30
14. Lockheed Lodestar 31
15. Taylorcraft Auster 32
16. Lockheed Ventura 33
17. Martin Maryland 34
18. Bristol Beaufort 35
19. Consolidated Catalina 36
20. Curtiss Tomahawk 37
21. Douglas Boston 38
22. Gloster Gladiator 39
23. Hawker Hurricane 40
24. Curtiss Kittyhawk 41
25. Curtiss Mohawk 42
26. Fairey Battle 43
27. North American Harvard ... 44
28. Bristol Blenheim 45
29. de Havilland Mosquito 46
30. Douglas DC-3/C-47 Dakota 47
31. Martin Baltimore 48
32. Bristol Beaufighter 49
33. Consolidated B-24 Liberator 50
34. Martin Marauder 51
35. Supermarine Spitfire 52
36. Vickers Wellington 53
37. de Havilland Dove/Devon 54
38. Douglas DC-4 55
39. Sikorsky S-51 Dragonfly 56
40. Short Sunderland 57
41. de Havilland Heron 58
42. de Havilland Vampire 59
43. Vickers Viscount 60
44. de Havilland Canada Chipmunk 61
45. North American Mustang 62
46. Avro Shackleton 63
47. Canadair/North American Sabre 64
48. English Electric Canberra 65
49. Sikorsky S-55 66
50. Sud Aviation Alouette II 67
51. Sud Aviation/Aérospatiale Alouette III 68
52. Cessna 185 69
53. Dassault Mirage III 70
54. Lockheed C-130 Hercules 71
55. Westland Wasp 72
56. Hawker Siddeley Buccaneer 73
57. Sud Aviation/Aérospatiale Super Frelon 74
58. Aermacchi/Atlas Impala 75
59. Sud Aviation/Aérospatiale Puma 76
60. Piaggio Albatross 77
61. Transall C-160 78
62. Aermacchi Bosbok 79
63. Hawker Siddeley Mercurius 80
64. Atlas Kudu 81
65. Dassault Mirage F1 82
66. Swearingen Merlin 83

Chapter 3 Squadrons 84

Chapter 4 Into the Future 93

Introduction

'There is absolutely no limit to the scale of its future independent war use. And the day may not be far off when aerial operations, with their devastation of enemy lands and destruction of industries and populous centres on a vast scale, may become one of the principal operations of war to which the older forms of military and naval operations may become secondary and subservient.'

General Jan Smuts

In 1917, the South African General Jan Smuts completed a review of British Air Services, the so-called 'Smuts Report', which contributed substantially to Great Britain's decision to set up a single air service in Britain. Prime Minister Lloyd George reckoned Smuts to be 'the father of the RAF'.

It would, however, be World War Two that was the catalyst for the creation of South Africa's air defence arm, albeit at a time when the nation was caught totally unprepared for such an eventuality. In terms of expertise, cooperation and training, the Royal Air Force (RAF) was a valuable partner in those formative years.

With the cessation of the war in 1945, the vast majority of what were essentially wartime squadrons were disbanded. However, driven largely by political policies that brought about the need for border counterinsurgency measures from the 1960s, almost all the mothballed squadrons were reactivated. Home-grown innovation witnessed the growth of an air force increasingly self-reliant in ordnance and technology.

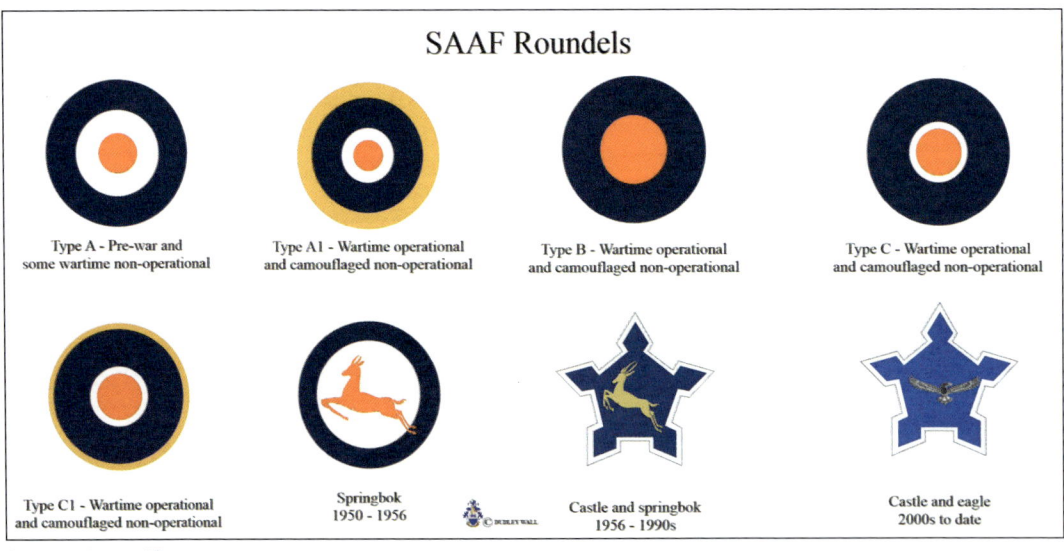

(Col Dudley Wall)

The South African Air Force (SAAF) was then in the transition to jet-powered aircraft, largely of French and Italian origin. This rapidly led to licensed local production of certain makes, thereby further reducing the nation's dependence on external sources. This development accelerated at a greater pace as the conflict in South West Africa, now Namibia, and Angola, intensified – the so-called Border War – in which the South African Defence Force was pitted against sophisticated Eastern-bloc weaponry.

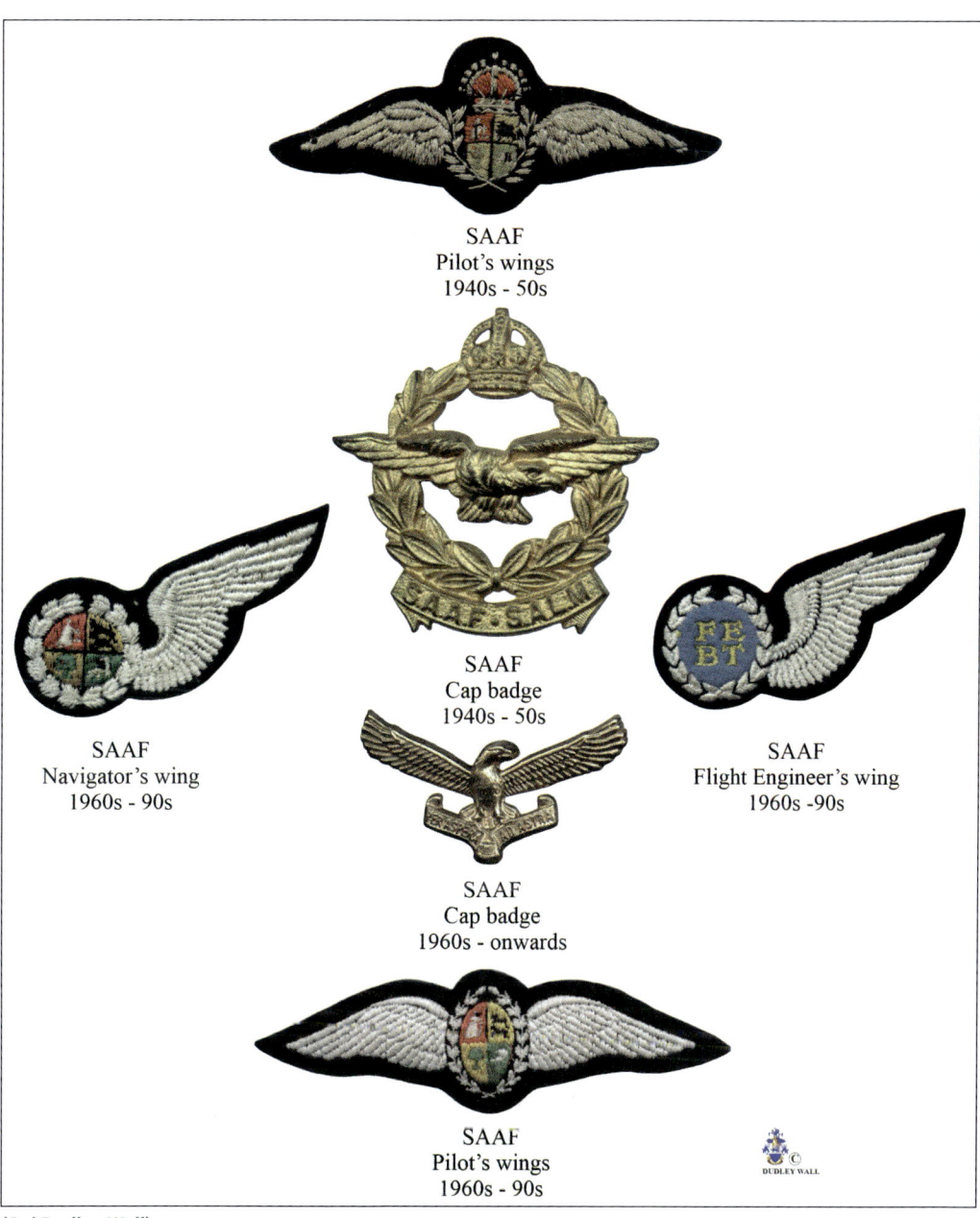

(Col Dudley Wall)

Chapter 1

History

During World War One, 3,000 South African pilots served in the Royal Flying Corps the flight arm of the British Army, effectively yielding a solid pool of pilots for the South African Air Force (SAAF), which was formed in 1920. This landmark development in the nation's armed forces was significantly bolstered by an 'imperial gift' from the British government worth two million pounds. The 113 aircraft included in the donation gave the fledgling air force a flying start.

However, at the time the South African Treasury did not share General Smuts's vision, and the embryonic SAAF floundered for want of funding. With such constraints, it would be 1932 before the first SAAF flying school was established and, suffice to say, at the outbreak of World War Two, South Africa's air service was wholly unprepared for active service. The SAAF's permanent strength stood at a meagre 160 officers, 35 officer cadets and 1,400 other ranks. Following intensive expanded training programmes at Pretoria, Germiston, Bloemfontein, Kimberley and Baragwanath, by the end of 1941, the SAAF's strength stood at more than 31,000, which included 956 pilots, 715 observers and gunners, and 2,940 basic trainees.

In his exclusive memoirs, compiled and self-published by the author, former SAAF wartime pilot, the late Lieutenant General Keith Coster SASS, ICD, OBE, reveals the accelerated pace at which the SAAF forced itself onto a war footing:

The de Havilland Tiger Moth II was used extensively by Commonwealth countries for pilot training during World War Two. (Gerry van Tonder)

When the war broke out I was an officer-cadet at the South African Military College. We were training to be officers in the South African Permanent Force, and whilst our training included infantry, artillery and air force subjects, we could only become either artillery or air force officers if we passed the course and were commissioned as officers.

The exponential growth in asset acquisition was as remarkable. In the same period, operational and training aircraft had swelled from 219 to 1,709. Tiger Moths were employed for initial training, and Ansons and Oxfords for type conversion and navigation and air gunnery training. From 1942, advanced training was conducted on Harvards, replacing the Harts and Masters.

We officer cadets had been out training on the day South Africa declared war, and we were summoned to return to the military college forthwith. We were told to go into one of the lecture rooms and sit down as the commandant of the college was going to address us. In came Lt Colonel William Henry Evered Poole, accompanied by the Adjutant Capt F.J. Dyason. The colonel addressed us as follows, 'I have two things to tell you today. Firstly, this morning the Royal Air Force bombed the Kiel Canal in Germany, and secondly, you are all commissioned as from today.'

At that stage we were all very close to getting our 'wings' for flying, and for the next ten days we put in quite a lot of flying time. On 16 September, I did my flying test with Major S.A. Melville and qualified for my wings, which meant that I was considered capable of carrying passengers in the air.

On 18 September 1939, I, with a number of other young officers, caught the train to Cape Town where we would join No. 5 Bomber/Fighter Squadron. From Monday to Saturday (we worked on Saturdays then), after an early breakfast, we were taken in a 3-tonner from the Wynberg officers' mess down to an aerodrome known as Young's Field in the Wetton area near the Kenilworth racecourse, where our flying duties started at 6.00a.m. Our first task was to convert to the Westland Wapiti, which was the service aircraft with which No. 5 Bomber/Fighter Squadron was equipped.

In order to get our 'wings', we had flown Avro Tutor biplanes, on which I went solo for the first time on 12 October 1939, after 11 hours and 50 minutes of flying instruction. I was then 18½ years old. Following our *ab initio* flying instruction, we later graduated to Hawker Hart biplanes and then the Hawker Hartebees, which were similar to the Harts, but more powerful. After 113 hours and 40 minutes of flying time, I qualified for my 'wings'.

The Supermarine Spitfire Mk V was flown by SAAF No. 40 Squadron in the North African Tunisian front 1943. (Col André Kritzinger, CC BY-SA 3.0, Wikimedia Commons)

During this period, the growth in the SAAF's asset acquisition was as remarkable. From a total operational and training aircraft strength of 219 in September 1940, a year later the number has swelled to 1,709.

Meanwhile, the SAAF had immediately commenced operational duties, employing former South African Airways Junkers Ju 86s on coastal reconnaissance and defence patrols. In due course, these would be replaced with Ansons, Marylands, Venturas and Catalinas.

As early as May 1940, the SAAF entered the East Africa theatre against Italian forces in Abyssinia, Somaliland and Eritrea. This consisted of SAAF No. 1 Squadron, flying Hurricanes and Furies, and Nos. 11 and 12 Squadrons flying Ju 86s and Hartebees in bomber roles. Outnumbered by the *Regia Aeronautica Italiana* (Royal Italian Air Force), this initial deployment was soon joined by SAAF No. 3 Squadron, flying Hurricanes, and Nos. 40 and 42 Squadrons, equipped with Hartebees. SAAF No. 11 Squadron, now re-equipped with Fairey Battles, significantly enhanced Springbok air power in the region.

As the successful East African Campaign wound down, South African air and ground forces joined the Allies in the Western Desert against the German–Italian *Panzerarmee Afrika*.

Arriving in April 1941, SAAF Nos. 1 and 24 Squadrons, flying Hurricanes and Marylands respectively, took part in Lieutenant General Archibald Wavell's 'great westward offensive', the last Allied victory of any consequence until the Second Battle of El Alamein in late 1942.

By September, No. 3 (South Africa) Wing had been formed, incorporating SAAF Nos. 12 and 24 Squadrons flying Marylands. Soon thereafter, and also flying Marylands, SAAF No. 21 Squadron joined the wing. In due course, the wing was re-equipped with Bostons and Baltimores. SAAF Nos. 1 and 2 Squadrons (fighter), No. 40 Squadron (tactical reconnaissance) and No. 40 Squadron (aerial reconnaissance) also joined the SAAF's Desert Air Force contingent.

Events in North Africa from the summer of 1942 were seminal for the SAAF, which many contend was the SAAF's 'finest hour'. Field Marshal Erwin Rommel's relentless eastward drive on Egypt and the Middle East beyond, pushed Allied ground and air forces to the brink of disaster. In July alone, SAAF fighters flew 2,344 sorties, and between 31 August and 4 September, SAAF Boston bombers flew 334 missions. In the second half of October, SAAF Nos. 1 (Hurricanes), 2 and 4 (Kittyhawks), and 5 Squadrons (Tomahawks) carried out 1,377 fighter and fighter-bomber sorties; an average of 106 a day. The SAAF continued with exhaustive air support right up to the end of the North Africa Campaign in Tunisia in May 1943.

With the ensuing Allied invasion of Sicily and mainland Italy, SAAF No. 1 Squadron flew bomber escort from Malta, while SAAF No. 40 Squadron provided air observation for the landing ground troops.

As the Luftwaffe's presence in the air over Italy waned dramatically and the SAAF's strength increased, in October 1943 a top-level meeting in London made a key strategic decision to redeploy RAF crews to Europe for a pre-invasion softening-up. This resulted in the formation of a heavy-bomber wing consisting of SAAF Nos. 31 and 34 Squadrons equipped with Liberators. SAAF No. 30 Squadron was formed as a light-bomber unit flying Marauders. In addition, three more Spitfire and Kittyhawk fighter and fighter-bomber squadrons were formed. Three coastal squadrons were deployed from South Africa to the Mediterranean, and RAF No. 262 Squadron, a Catalina-flying unit, was taken over by the SAAF (retitled No. 35 Squadron in 1945).

In the unforgiving Allied campaign that relentlessly and painfully pushed Field Marshal Albert Kesselring's Army Group C through the Northern Apennines and ultimate surrender, the SAAF experienced some of its most active times of World War Two. Between the Axis capitulation in North Africa and the German surrender in Italy, SAAF squadrons flew a total of 82,401 sorties. Its strength peaked at 35 squadrons, of which 27 were active in the Italian Campaign. During the five years of the war, the SAAF lost 968 aircrew killed and 320 missing.

With the end of the war, many squadrons were disbanded and the SAAF re-organised. (see Chapter 3). In 1946 and 1947, the remaining squadrons performed locust and tsetse fly control at home and in Kenya.

SAAF No. 2 Squadron Canadair F-86F (CL-13B) Sabres were pitted against Soviet MiG-15s during the Korean War, 1953. (Ad Meskens, CC BY-SA 4.0, Wikimedia Commons)

However, the emergence of global Cold War crises quickly put an end to this period of relative calm. In 1948, 20 SAAF aircrews were despatched to West Germany to assist with relief operations necessitated by the Soviet blockade of Berlin – the Berlin Airlift.

Then, in September 1950, SAAF No. 2 Squadron, the 'Flying Cheetahs', joined American-led United Nations forces combating the Soviet-sponsored invasion of South Korea by North Korean forces. Converting to the Mustang F-51 at Johnson (US) Air Force Base, the Springboks were attached to the US 18th Fighter-Bomber Wing, flying out of the occupied North Korean capital, Pyongyang. Two months later, following the Communist Chinese push down the Korean peninsula, the squadron was relocated to US K-10 airbase at Chinhae on the southern tip of Korea.

Its home for the next two years, Chinhae saw No. 2 Squadron fly 10,373 armed reconnaissance, interdiction and ground-support sorties, losing 12 pilots killed and 74 of its 95 aircraft. Early in 1953, the squadron converted to F-86 Sabre jets, and in March moved to US K-55 airbase at Osan, 22 miles south of Seoul.

Here, until the armistice on 27 July that year, the squadron flew 2,023 Sabre sorties. In a unique event in the history of the SAAF, in August 1956, No. 2 Squadron was awarded a US Presidential Citation for 'extraordinary heroism in action against the armed enemy of the United Nations'.

For the remainder of the 1950s and the next two decades, plans to acquire new aircraft were implemented, including the purchase of Canberras, Dassault Mirage IIIs, helicopters and Buccaneers.

As the international United Nations-endorsed arms embargo on South Africa started to bite, licensed aircraft production commenced in the country with the establishment of the Atlas Aircraft Corporation. During this period, the facility would build the Aermacchi MB-326K, locally dubbed the 'Impala', and assemble Dassault Mirage F1-AZ strike aircraft.

Strike Command and Air Transport Command had their headquarters at AFB Waterkloof near Pretoria, while Maritime Command was based at Cape Town's D.F. Malan Airport. A strong and highly versatile helicopter element arose out of the purchase of predominantly Aérospatiale aircraft from France, including Pumas, Super Frelons and Alouette IIs and IIIs.

The iconic French delta-winged Dassault Mirage III first entered SAAF service in 1963 with the arrival of five disassembled aircraft in the cargo hold of a C-130 Hercules. (SAAF)

Harvards for flying training were gradually replaced by Impalas, while advanced flying schools employed Sabres, Mirage IIIs and Dakotas. The country's Active Citizen Force continued to play a key role in both ground and air defences, with 13 air commando squadrons raised under the control of the SAAF and equipped with civil light aircraft for use in emergencies. By the time the SAAF celebrated its 50th anniversary in 1970, it had become a powerful force on the continent.

In the mid-1970s, hitherto relatively isolated nationalist guerrilla incursions from Angola into South African-administered South West Africa rapidly developed into what the South Africans referred to as the Border War. The conflict was characterised by South African cross-border attacks into Angola against Soviet- and Cuban-backed government forces which had thrown their weight behind the guerrillas.

In this conflict, the SAAF was a major player, in some cases testing its attack and logistical capabilities to the limit. During Operation *Savannah* in 1975, this included the long-distance bombing by SAAF No. 12 Squadron Canberras of targets deep inside the Angolan hinterland.

In May 1978, the Operation *Reindeer* air-and-ground assault on Cassinga in southern Angola took place, in which Canberras and Buccaneers softened up the target in preparation for C-130 and C-160 para drops. Mirage IIIs from AFB Waterkloof provided essential air cover and support. Helicopters, particularly the Puma and the Alouette III, played an essential air-supply and medical-evacuation role.

After the Angola campaign, the SAAF remained fully engaged on the border until 1989, when South Africa's defence forces returned home in anticipation of South West Africa becoming the independent state of Namibia.

Affectionately dubbed the 'Samaritan with teeth' by the South African Defence Forces, the extremely versatile French-made Aerospatiale Puma flew anything from operational troop deployment missions to medical evacuations (casevacs) during the Border War. (Col André Kritzinger, CC BY-SA 3.0, Wikimedia Commons)

Mirage IIIRZ. The origins of the shark-mouth nose art, popularised by the AVG Flying Tigers' P-40 Warhawks in World War Two, can be traced back to German LFG Roland C.II reconnaissance biplanes of World War One. (Aeroprints, CC BY-SA 3.0, Wikimedia Commons)

Chapter 2
Aircraft

With the outbreak of World War Two in September 1939, spare aircraft to sell among the Commonwealth nations were in very short supply. Those previously purchased or built in South Africa were obsolete, with only six Hawker Hurricane Mk 1s, a Fairey Battle and a Blenheim Mk 1 considered suitable for operations. Added to this, the 1936 plan for expansion was yet to be implemented.

However, the requisition of the entire South African Airways passenger fleet of German-made Junkers aircraft fulfilled an immediate need for transports and light bombers. Under the Peace Expansion Scheme, in October that year a total of 720 aircraft were acquired, of which 336 were fighters.

While the Tiger Moth became the workhorse at nationwide training schools, Ansons and Oxfords were relied on for twin-engine conversions and air gunnery training. The North American T-6, referred to locally as the Harvard, was acquired for advanced training.

The campaign in Italian East Africa would be a baptism of fire for the SAAF. In May 1940, SAAF No. 1 Squadron entered the fray in its Hawker Hurricanes (its most modern fighter at the time) and Furies. Soon after, they were joined by Gladiators, Hartebees and Ju-86 bombers. Though outdated, the locally produced Hartebees, which provided the bulk of the SAAF's serviceable aircraft, proved their worth in combat in the arid, semi-desert conditions of the region.

In due course, Fairey Battles replaced the Ju 86s, and Marylands entered the conflict in a reconnaissance and survey role. With the defeat of the Italians in East Africa, the Marylands were joined by Beauforts operating in ground support and reconnaissance roles during the Allied invasion of Vichy French-held Madagascar in May 1942.

Initially equipped with Hurricanes and Marylands, the SAAF moved to the Western Desert theatre in April 1941. Shortly thereafter, the Marylands were replaced with Bostons and Baltimores.

Taken over by the SAAF from the national airline, South African Airways, the Junkers Ju 52s had originally been built in Germany to serve Lufthansa. (Markus Kress, CC BY-SA 3.0, Wikimedia Commons)

After several setbacks, during 1942 the fortunes of war turned in favour of the Allies in North Africa. SAAF Kittyhawks and Tomahawks joined the Hurricanes escorting Bostons and Baltimores on unceasing bombing raids on Axis targets.

Arriving in Egypt from South Africa in April 1942, SAAF No. 5 Squadron pilot, Captain Keith Coster, was sent on a conversion course at No. 71 Fighter Operational Training Unit at Carthago airfield, where,

We flew Harvards to get the hang of flying single-engine, low-wing monoplanes with retractable undercarriages and variable pitch propellers, and graduated from these on to Hurricanes, and finally the American Tomahawks.

The Hurricane was an exciting and delightful aeroplane to fly. The Tomahawk was heavier and less manoeuvrable but a bit faster. Apart from becoming familiar with the characteristics of the aircraft, we were trained in formation flying, air gunnery – both air-to-air and air-to-ground – and dive bombing, as well as in the basic principles of air-to-air combat. The course finished on the 18 June 1942, by which stage I had clocked up 11½ hours on Harvards, 12¼ hours on Hurricanes and 17¼ hours on Tomahawks. I was assessed as 'Above Average' as a fighter pilot, 'Above Average' as a pilot-navigator, and 'Average' in air gunnery.

On 11 July, Coster was assigned to a medium-cover operation of 18 Bostons. Flying at about 8,000 feet, suddenly the escorts were ordered over the radio to attack a formation of German Stukas that were dive-bombing Allied forces in the same general area that the Bostons were preparing to bomb. The SAAF No. 5 Squadron formation broke away from the Bostons and dived in pursuit of the Stukas. Coster relates,

I had my sights on a Stuka and I fired my four .50 Browning guns at him. I know I damaged the Stuka, but whether he went down or not I don't know, because I became pre-occupied with a sudden loss of my ring-sight that normally imaged on the windscreen in front of my face. We all pulled out of our dive as we were getting dangerously close to the ground, and as I levelled out, there was an almighty bang. I realised that I had been hit!

Several SAAF squadrons flew the American Curtiss P-40 Warhawk, called the Tomahawk by the Commonwealth countries. (TMWolf, CC BY-SA 2.0, Wikimedia Commons)

While I was absorbing this development, a [Messerschmitt] 109 passed in front of me and I gave him a burst from my Brownings – but it was not an aimed burst as my ring-sight had gone. In trying to follow the 109, it became immediately obvious that my rudder had been shot away because there was no response when I kicked my rudder bar. An aircraft without a rudder becomes a sitting duck, as the basic manoeuvre in aerial combat is to turn into any aircraft that fires at you.

It was obvious that without rudder control I couldn't make any further contribution to the dogfight, so by the use of ailerons alone, I managed to start a long flat turn towards the sea when I was hit again. Not being able to turn into my attacker, I had but two alternatives: to climb or to dive. I decided on the latter and dived down to a few hundred feet above the ground.

After his Tomahawk was shot down in the Western Desert in 1942, SAAF Captain Coster was taken prisoner. (Gerry van Tonder)

In the dive I was hit a third time from directly behind – a long burst which started my Tomahawk burning and caused shrapnel wounds in my left arm and in my neck. I could see the long trail of smoke tailing out behind my aircraft and I knew that I had to make a very quick decision. To climb to a height where I could bail out would be inviting further – and probably fatal – attacks, or a complete burnout of the aircraft.

As I was very close to the ground, I decided to put it down on the desert and get out before it went up in flames. I loosened my straps and jumped clear before the aircraft had come to a halt. When I stopped rolling, I jumped to my feet and ran to put as much distance as I could between me and the aeroplane. Seconds later, it burst into flames and very rapidly burnt out completely.

Captain Coster was taken prisoner and spent the rest of the war in numerous concentration camps in Italy and Germany, including the famous Stalag Luft III.

With the imminent capitulation of Axis forces in North Africa, SAAF Mosquitoes conducted large-scale photographic surveys of parts of Sicily in preparation for Operation *Husky*, the Allied invasion of the Italian island. While the Germans withdrew to the Italian mainland, a detachment of SAAF Spitfires was deployed to the island of Cos in the Italian Dodecanese.

SAAF Marauders assisted with the difficult task of providing cover for the Allied Italian beachheads at Salerno and Anzio. This marked the start of significant growth of the SAAF's strength in Italy, arising out of the British Air Ministry's allocation to the Springboks of increased commitments in the Mediterranean. A Liberator heavy-bomber wing, three Spitfire and Kittyhawk squadrons and a second Dakota transport squadron were raised. At the same time, SAAF Catalinas brought up from home took over RAF No. 262 Squadron. Farther afield, SAAF Wellingtons provided top cover for Allied shipping during the Battle of the Atlantic.

AFB Ysterplaat gate guard Impala Mk I (Aermacchi MB-326) No. 531. (Oleg V. Belyakov, CC BY-SA 3.0, Wikimedia Commons)

The post-war years saw the SAAF continuing to play its part in the international arena. SAAF crews operated RAF Dakotas in the Berlin Airlift from 1948, and in 1950 SAAF No. 2 Squadron, flying Mustangs and then Sabres, joined UN forces in the Korean conflict.

The SAAF's Spitfires, Venturas and Sunderlands were gradually retired in the 1950s, while the Vampire, the SAAF's first jet fighter, was delivered in 1950. Other new aircraft brought into service in this decade included Shackletons, MR3s and Sabres.

In the 1960s, the SAAF entered an upgrading phase, acquiring Dassault Mirage III, Canberra, Buccaneer and C-130 Hercules aircraft. The rotary-wing element was also modernised with the acquisition of Alouette, Puma, Super Frelon and Wasp helicopters.

An important development at this time was the decision to both acquire and manufacture under licence in South Africa the Italian Aermacchi MB 326, locally renamed the Impala.

The SAAF's transport needs were also addressed, as seen by the acquisition of the Hercules C-130 'Flossie' in 1963 and the Transall C-160 in 1969. In the reconnaissance/light transport role, Cessna 185s, Aermacchi Bosboks and locally produced Kudus were introduced.

Not since World War Two had the SAAF faced such enormous combat challenges as during the 1966–89 Border War in northern South West Africa (now Namibia) and southern Angola. The vast majority of the SAAF's aircraft types were put to the test in every aspect of modern warfare against an enemy supported by Soviet and Cuban technology and manpower.

In May 1978, South African forces launched a massive airborne raid against a Namibian guerrilla base in Cassinga, Angola. In Hanlie Smith and Gerry van Tonder's *North of the Red Line: Recollections of the Border War by Members of the South African Armed Forces: 1966–1989* SAAF navigator Floris Brand gives an eyewitness account of zero hour, revealing the complex interaction between elements of the SAAF, all of which demanded meticulous timing:

> I was fortunate to have a ringside seat of the Canberras, Buccs [Buccaneers] and Mirages attacking Cassinga. I was second navigator on the Transie [Transall C-160], commanded by Commandant Steenkamp, which dropped the stopper group from east to west on the southern side of the target. As we were climbing from low level (LL) to 600 feet for the drop, the Buccs crossed directly underneath us one minute before the target. They were still dropping 30 seconds before we arrived.
>
> As the first navigator, Oom Louis Nel had taken over the drop at that time; I had just a moment to watch the attack unfold. Sitting at 600 feet, we saw first the Cans [Canberras] dropping from north to south, followed by the Buccs just to our right do their thing. We dropped under the canopy of the mushroom. I am sure our starboard wing went through the column.
>
> Viewing the strike from a vantage position left a lasting impression on me. I remember tracers coming at us from the target and making myself small behind the starboard window. We had a problem closing the starboard paratrooper door after the drop. I believe the loady [loadmaster] wanted some of the action, so we extended the run a bit to get the door closed before applying serious aerodynamics to the airframe. All the while, Mirages were attacking the target and going out underneath us to the south.

The SAAF aircraft looked at in this chapter cover the period from World War Two until the Border War; a time of dramatic growth interlaced with momentous challenges. Suffice to say, therefore, the majority are no longer in service, and references to squadrons and air force bases are applicable to this period.

In the post-World War Two years, South Africa established a lasting relationship with French aerospace manufacturers Dassault and Aérospatiale for a wide range of military aircraft and helicopters, such as the Mirage F1 (above) (clipperarctic, CC BY-SA 2.0, Wikimedia Commons) and SA330 (below). (Eric Salard, CC BY-SA 2.0, Wikimedia Commons)

1. de Havilland Tiger Moth

Left: The DH.82 Tiger Moth was the backbone of Commonwealth flying training during World War Two. About 700 Tiger Moths, sourced from Britain and Australia, saw service in seven training schools in South Africa. (Bob Adams, CC BY-SA 2.0, Wikimedia Commons)

Middle: Described as a two-seat elementary trainer and communications aircraft, the British-built DH.82 Tiger Moth was a reliable biplane built to withstand rough handling. Powered by a Gipsy Major inline piston engine, the Tiger Moth had a maximum speed of 93mph and a range of 302 miles. Generally docile in normal flight, the Tiger Moth responds well to control inputs, and is reasonably easy to fly for a 'tail-dragger'. Its large 'parachute' wings are very forgiving, and it stalls at a speed as slow as 25 knots with power. (Gerry van Tonder)

Below: The original 52 Tiger Moths obtained by the SAAF were civilian aircraft that were taken over for training purposes at the start of World War Two. They initially had civilian paint schemes, including a red fuselage with silver wings. They were soon painted in an all-over trainer yellow. (Col Dudley Wall)

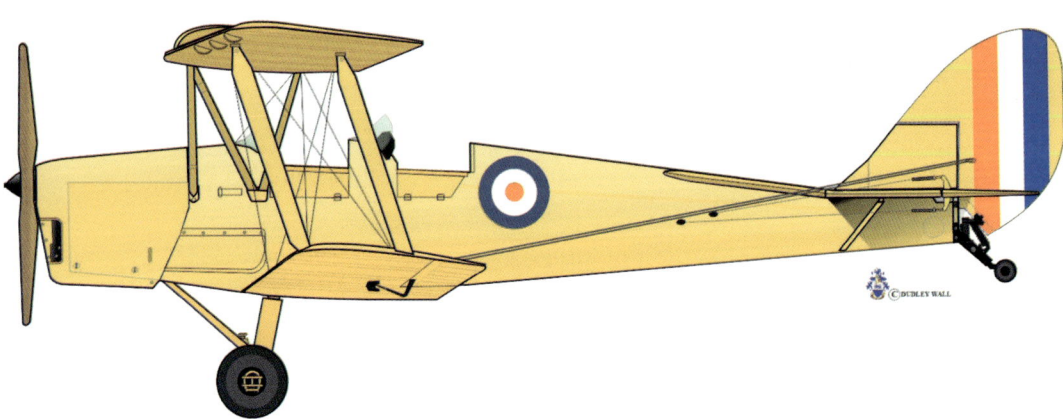

2. Hawker Fury

Powered by a Rolls-Royce Kestrel piston engine, the British Hawker Fury was armed with a forward-firing .303 machine gun and a Lewis gun in the aft cockpit. Underwing hardpoints could accommodate two 112lb bombs. (Alan Wilson, CC BY-SA 4.0, Wikimedia Commons)

In 1935, South Africa purchased seven Furies, which were delivered in September 1936. These were the first single-seat fighters for the SAAF since the SE.5As received as part of the Imperial gift in 1920. Early in World War Two, Furies of SAAF Nos. 1 and 2 Squadrons saw action in East Africa against Italian forces. Their last significant use was with SAAF No. 43 Squadron in the 'Air Commando' tour of South Africa, when air displays were performed for recruiting and propaganda purposes. (Col Dudley Wall)

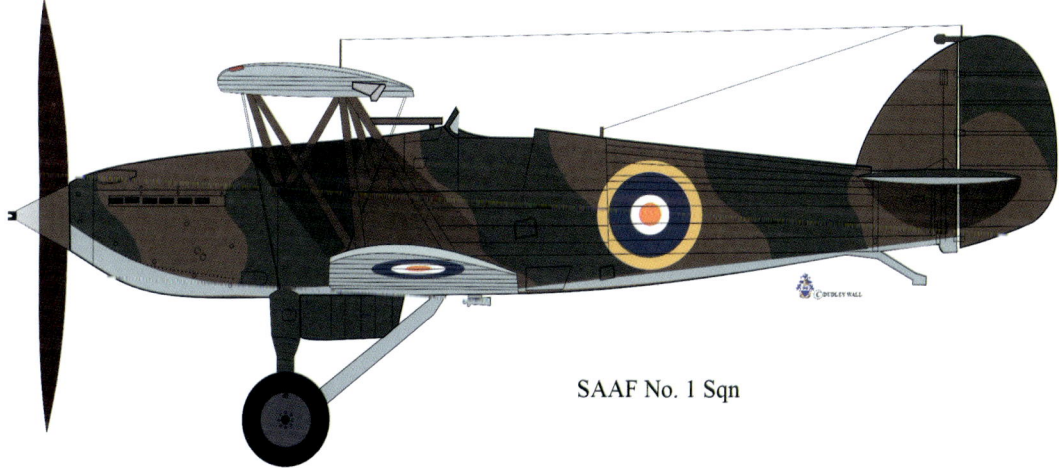

SAAF No. 1 Sqn

3. Bücker Jungmann

The German Bücker Bü 131 Jungmann ('young man') was a basic 1930s two-seater training aircraft primarily used by the Luftwaffe during World War Two. Powered by a Hirth HM 60R four-cylinder, inverted inline engine, the Jungmann had a top speed of 114mph. The Jungmann was a tubular steel and wood construction, covered in fabric, with the exception of the frontmost fuselage segment, which was covered with sheet metal. First taking to the air in April 1934, the aircraft was prized for its outstanding handling characteristics when compared to other contemporary biplanes. It was exceptionally sturdy and agile with two open cockpits in tandem and fixed landing gear, features which created substantial demand in the 1930s. The military air services of Imperial Japan and the Kingdom of Yugoslavia became the largest export customers, acquiring some 1,300 and 400 aircraft respectively, while hundreds were manufactured under licence by the Spanish aircraft company Construcciones Aeronáuticas SA (CASA). After the war, the Jungmann continued to be flown by several operators, including the Spanish Air Force, where the aircraft remained the primary basic trainer right up to 1968. (Col Dudley Wall)

By 1938, 16 Jungmanns had been privately imported into South Africa. At the start of World War Two, the SAAF requisitioned these civil aircraft for use to train pilots and staff on the repair and maintenance of aircraft. Many of these aircraft were also employed for communications between headquarters and the air schools and operational training units. (Crazy Jet)

4. Hawker Hart

In 1929, the Hart was the first of that long line of beautiful silver-winged biplanes that served the RAF throughout the 1930s. The performance of the inline engined 'all-metal construction' two-seat, light day bomber was such that this and related Hawker designs were to dominate RAF and several Commonwealth countries' equipment almost until the declaration of war. During RAF exercises in 1931, and with a maximum speed of 160mph, the Hart was considerably faster than contemporary RAF bombers or fighters. The Hart was armed with one fixed forward-firing .303 Vickers machine gun in the port side of the nose and a .303 Lewis gun on a rear cockpit mounting, plus a bomb load of up to 520lb. (Alan Wilson, CC BY-SA 4.0, Wikimedia Commons)

From 1937, 100 British-made Harts were purchased from the RAF by South Africa to serve as advanced trainers and second-line operational aircraft. The Hart bomber had a rear gun position whilst the Hart trainer had dual controls. Retired by the RAF in 1938, the Hart remained in service with the SAAF in a communications role until 1943. (SAAF)

5. Junkers Ju 52/3m

The German national carrier Lufthansa became one of the first customers of one of the longest-serving aircraft ever built. In the mid-1930s, the Junkers Ju 52/3m was taken on by the Luftwaffe as the backbone of its bomber and military transport strength. With its distinctive three BMW 132A radial piston engines, the Ju 52/3m had a reputation for durability and versatility. (Rschider, GFDL V 1.2, Wikimedia Commons)

During the East African Campaign (10 June 1940 to 27 November 1941), SAAF No. 50 Squadron operated a shuttle service between the theatre in East Africa and South Africa as part of No. 1 Bomber Transport Brigade. Operating Ju 52/3m aircraft requisitioned from South African Airways at the outbreak of war, 11 were assigned to SAAF No. 51 Flight, which operated a shuttle service from Nairobi to Egypt, the Middle East and South Africa. During the war, the SAAF flew 12 Ju 52/3ms captured in the North African theatre. One of these, 'Tobruk 26', was used by the SAAF for shuttle flights between South Africa and Egypt. SAAF No 19 Squadron, formed on 1 September 1939 with 17 Squadron to form part of the Airways Wing at AFB Swartkop, flew ex-SAA Junkers Ju 52/3ms. The squadron was disbanded on 1 December that year. In 1944, South African Airways restarted domestic air routes and the remaining Ju 52/3ms were put into storage until the late 1940s, when they were mostly sold or retired. (Col Dudley Wall)

6. Junkers Ju 86Z

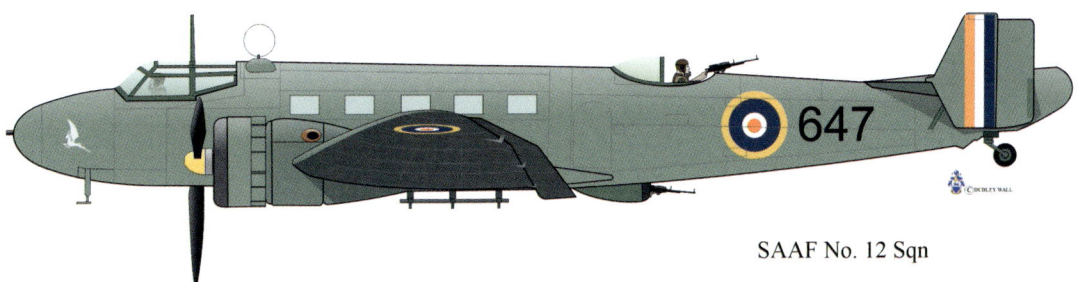

SAAF No. 12 Sqn

South African Airways ordered 17 German Junkers Ju 86Z and one Ju 86K airliners, the first arriving in June 1937. After the first five were produced with Rolls-Royce Kestrel V-12 engines, the rest were fitted with Pratt & Whitney R-1690 Hornet radial engines, and given the designation Ju 86Z-7. When World War Two broke out, the SAAF requisitioned the airliners as bombers and transports, initially with SAAF No. 15 Squadron at Wingfield, Cape Town, before moving to No. 12 Squadron. (Col Dudley Wall)

In mid-1940, SAAF No. 12 Squadron entered the Italian East Africa theatre with their Ju 86s. In May 1941, the remaining eight Ju 86s of No. 12 Squadron were transferred to newly formed SAAF No. 16 Squadron, seeing action during the Allied advance into Gimma in Italian-occupied Ethiopia. In August 1941, the unit ceased to operate as a squadron, and was renumbered No. 35 Flight. The SAAF retained two Ju 86s as VIP aircraft. The Ju 86 ZS-ANI, pictured below, was the only K-1 bomber variant delivered to South Africa. It was fitted with BMW engines. Named *Lady Ann Barnard* by South African Airways, when it was requisitioned by the SAAF it was fitted with turrets and bomb racks and simply numbered '658'. (JWM Model)

7. Avro Tutor

The British Avro 621 Tutor biplane was adopted by the RAF as its basic trainer in the early 1930s, replacing the Avro 504N. Initially powered by an Armstrong Siddeley Lynx IVC radial piston engine, the Tutor had a maximum speed of 122mph. (RuthAS, CC BY-SA 2.0, Wikimedia Commons)

After receiving a licence, South Africa ordered 49 Tutors, which were built at the Aircraft and Artillery Depot, Roberts Heights (Voortrekkerhoogte), Pretoria, for the SAAF. Deliveries began in 1935. By the end of the war, the SAAF's Tutor strength stood at 59. From 1942 to 1945, Tutors were used at SAAF No. 48 Air School, Woodbrook, East London, for elementary navigation training, and at SAAF No. 62 Air School, Tempe, Bloemfontein, for flying instructor training. (foundin_a_attic, CC BY-SA 2.0, Wikimedia Commons)

8. Hawker Audax

The Hawker Audax was a development of the Hart in the army co-operation role, with a reduced bomb load and a message pick-up hook added. A total of 82 were received by the SAAF for use with the Service Flying Training Schools. Armament comprised a fixed, forward-firing .303 machine gun and a Lewis gun in the rear cockpit. There were also underwing hardpoints for either four 20lb or two 112lb supply containers. (Col Dudley Wall)

Described as a two-seat light bomber, which was an early design attributed to Sydney Camm, the Hawker Audax entered service with the RAF in 1934. It was immediately apparent that the type had considerable potential for adaptation to a number of other roles, and as a result, Specification 7/31 was issued, seeking an aircraft to replace the Armstrong Whitworth Atlas. Initially, it was powered by the Rolls-Royce F.X1 water-cooled engine, which later became better known as the Rolls-Royce Kestrel IB. Around 400 RAF aircraft remained in service at the start of World War Two, flying operational sorties in East Africa, along the Kenya–Abyssinia border as well as from RAF Habbaniyah during the Iraqi uprising of May 1941. After 1939, ex-RAF aircraft were also supplied to India, South Africa and Southern Rhodesia. (Public domain)

9. Hawker Hartebees

The Hawker Hartebees was essentially a tropicalised variant of the Hawker Audax, designed to meet the SAAF's need for a ground-support aircraft, with the main adaptation being the substitution of the powerplant with the much stronger Rolls-Royce Kestrel VFP inline piston engine. Armament remained the same as that of the Audax. Widely described as either the 'Hartebeest' or 'Hartebees', the latter was the designation used on the aircraft drawings, contracts and production licence and is therefore the spelling used here. The Hawker Hartebees was procured against Specification 22/34, with four pattern aircraft (SAAF 801–804) being built by the renamed Hawker Aircraft Ltd. The first two aircraft were standard machines, whereas the other two featured increased armour for the aircrew. In mid-1940, the type was used operationally alongside the Hawker Hart. After the war, the surviving aircraft were used as trainers in Southern Rhodesia and South Africa. The type continued in SAAF service until at least 1946. A single example of the Hawker Hartbees (SAAF 851 – both images on this page) is preserved at the Museum of Military History, Saxonwold, Johannesburg. (BAe Systems)

From 1937, 65 Hartebees were built under licence at the Aircraft and Artillery Depot at Robert's Heights (Voortrekker Heights), near Pretoria. Supplied to SAAF Nos. 40 and 41 Squadrons, at the outbreak of World War Two, 53 remained in service. In mid-1940, the Hartebees saw considerable action against Italian forces on the Kenya-Ethiopia border during the East African Campaign. (Alan Wilson, CC BY-SA 4.0, Wikimedia Commons)

10. Hawker Hind

In 1934, the Hawker Hind T.Mk. I light bomber was developed to bridge the gap until the new-generation aircraft, such as the Bristol Blenheim and the Fairey Battle, began to enter RAF service. A new derivative of the Hawker Hart, adaptations included the more powerful Rolls-Royce Kestrel V engine, modifications to the rear cockpit to improve firing and bombing efficacy and the replacement of the tail skid with a tailwheel. (Gerry van Tonder)

The SAAF received a total of 128 Hawker Hind bomber and trainer variant aircraft, which saw service in the Union of South Africa over the period 1940 to 1944. During this period, the SAAF received a further 63 Hinds retired from the RAF. These were not allotted SAAF serial numbers, being recorded 'struck-off charge in the Union of South Africa.' (SAAF)

11. Westland Wapiti

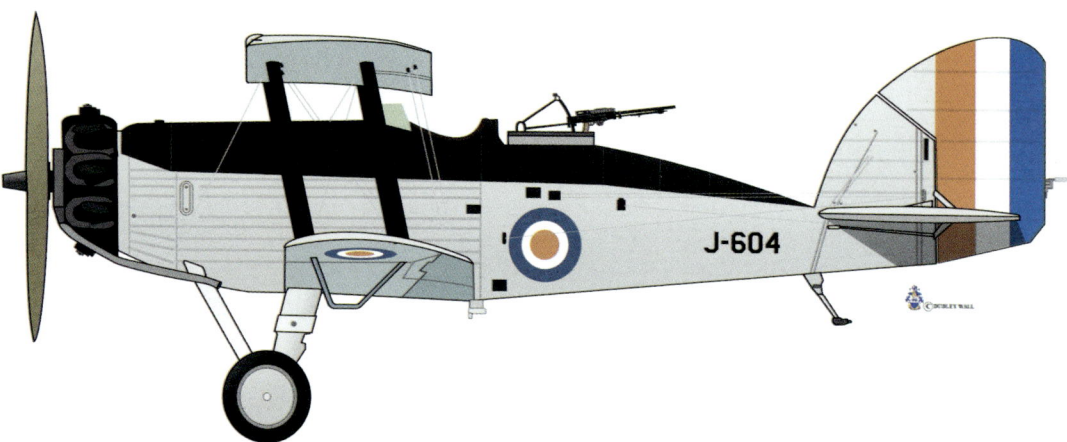

The Westland Wapiti was a two-seat, general-purpose, single-engined military biplane of the 1920s, designed and built by Westland Aircraft Works to replace the DH.9A in RAF service. The Wapiti was armed with a Vickers synchronised machine gun and a Lewis gun on a Scarff ring over the rear cockpit, plus up to 580lb of bombs. (Col Dudley Wall)

A total of 31 Wapitis served in the SAAF, of which 27 were the Mk III, which were licence-built in South Africa. SAAF No. 6 Squadron, formed in Cape Town in April 1939, was equipped with Westland Wapiti IIIs. Initial duties at the outbreak of war were that of anti-submarine coastal patrols from Youngsfield as part of Coastal Command SAAF. (SAAF)

12. Airspeed Oxford

The Airspeed AS.10 Oxford was first introduced to the RAF's Central Flying School in 1937. From August 1940, 665 aircraft (424 Mk I and 241 Mk II) were taken by the SAAF to train pilots on multi-engine aircraft under the Empire Air Training Scheme. Records reveal that 'attrition from accidents was notably high'. (Tony Hisgett, CC BY-SA 2.0, Wikimedia Commons)

Essentially a variant of the Airspeed Envoy, the Oxford had the same wooden construction, tailwheel-type retractable landing gear and basic airframe. In addition to the pupil and instructor seats, there were positions for the training of an air gunner, bomb aimer, camera operator, navigator and radio operator. (Mike Freer, GFDL V 1.2, Wikimedia Commons)

13. Avro Anson

One of the longest production runs of any British aircraft, between 1935 and 1952, saw some 11,000 Avro 652As come off the assembly line. In an operational role, the Anson's armament included one forward-firing .303 machine gun and one in the dorsal turret. It could carry up to 360lb of bombs. (Oren Rozen, CC BY-SA 2.0, Wikimedia Commons)

Over 700 GR1 and T1 Ansons were shipped to South Africa, to be used on maritime duties and for navigation, bombing, gunnery and radio training. Under the Empire Air Training Scheme, the SAAF established seven flying training groups, resulting in the Anson rapidly becoming one of the principal trainer aircraft in the SAAF. Ansons were also used in East Africa and the Middle East. After the war, they remained in use as trainers until the early 1950s. (Alan Wilson, CC BY-SA 4.0, Wikimedia Commons)

14. Lockheed Lodestar

In the 1930s, South African Airways (SAA) purchased 29 American-built Model 18.08 Lockheed Lodestar passenger transport aircraft from new. Powered by two Pratt & Whitney Hornet radial piston engines, the Lodestar could take up to 14 passengers. Lockheed, however, was unable to compete with the universally successful Douglas DC3/C-47, which became the predominant American transport aircraft of World War Two. (NJR ZA, Wikimedia Commons)

South African Airways (SAA) Lodestars were commandeered by the SAAF for wartime service. South African Prime Minister, Field Marshal Jan Smuts, used an ex-SAA Lockheed Lodestar, named *Jan van Riebeeck*, as his personal transport aircraft to tour the North African theatre (above). SAAF No. 50 Squadron flew Lodestars out of AFB Swartkop, East London and Broken Hill before being absorbed into Bomber Transport Command at Germiston in December 1940. (SAAF)

15. Taylorcraft Auster

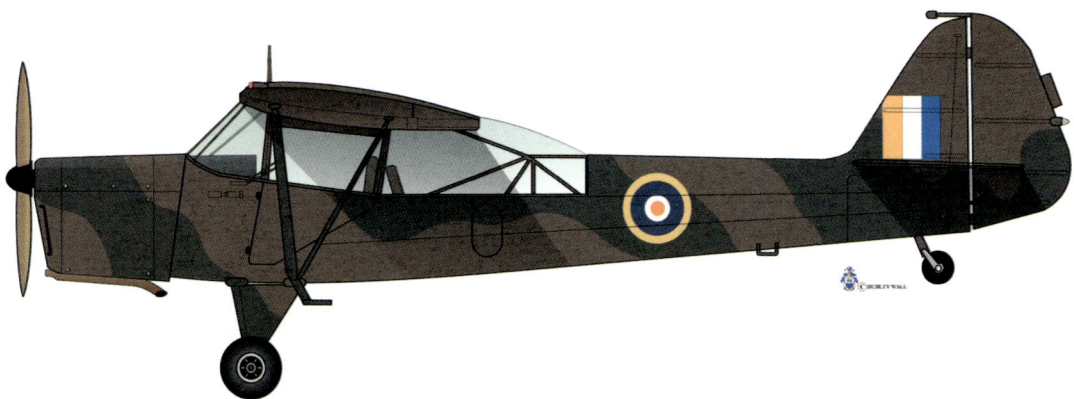

Of American origin, the Taylorcraft Auster was a British-built military liaison and observation aircraft, constructed of fabric-covered tubular steel with wings of wooden spars and aluminium ribs. (Col Dudley Wall)

After the end of World War Two, six ex-RAF Auster Mk. Vs were sent to South Africa where they served with 42 AOP Flight (later No. 42 Squadron) at Dunnottar training Air Observation Post pilots, and at Potchefstroom, training in cooperation with artillery. These pilots became known as 'tekkie pilots', because of the trainer-type footwear they were forced to adopt, due to army boots not coping well with the tiny heel-brakes of these small aircraft. These SAAF Mk Vs probably set a record for the longest-serving military Austers in the world, as they were only withdrawn in 1962. (Alan Wilson, CC BY-SA 4.0, Wikimedia Commons)

16. Lockheed Ventura

The American Lockheed Ventura was a twin-engine medium bomber and patrol aircraft, which first entered combat in Europe as a bomber with the RAF in late 1942. (SAAF)

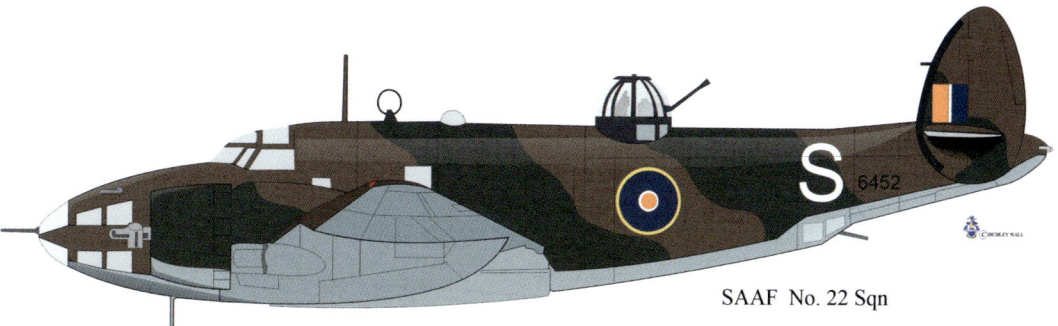

SAAF No. 22 Sqn

The SAAF received some 135 PV-1 Ventura GR-5s to protect shipping around the Cape of Good Hope. Venturas served in the SAAF until 1960. (Col Dudley Wall)

From 1943, the Ventura saw active war service with SAAF No. 17 Squadron in the Middle East, Egypt, North Africa and Sardinia, and with SAAF No. 22 Squadron on anti-submarine operations from Gibraltar. Home maritime patrol and reconnaissance/bomber duties were also performed by Venturas in SAAF Nos. 24, 25, 26, 27 and 28 Squadrons. (Ron Belling 1954, SAAF)

17. Martin Maryland

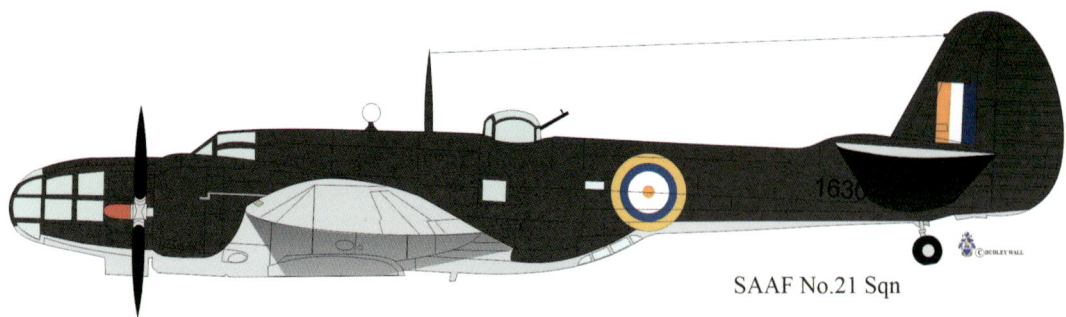

SAAF No.21 Sqn

The American Martin Maryland was a three-seat reconnaissance/bomber aircraft, powered by two Pratt and Whitney R-1830 Twin Wasp radial piston engines. Its armament included four .303 wing-mounted Browning machine guns, a .303 Vickers K gun each in dorsal and ventral positions, plus a bomb load of up to 2,000lb. (Col Dudley Wall)

The SAAF took delivery of 82 Maryland Mk IIs. The Marylands of SAAF No. 16 Squadron first saw combat during Operation *Ironclad*, the successful Allied invasion of Madagascar in 1942. Those of SAAF No. 21 Squadron served in the East African Campaign against Italian forces in Ethiopia, and later in the North African theatre until November 1943, when they were replaced with more modern Martin Baltimores. Marylands also saw service briefly with SAAF No. 24 Squadron in North Africa. (RAF)

18. Bristol Beaufort

SAAF No.16 Sqn

Responding to the British Air Ministry's requirement for a torpedo and reconnaissance bomber, Bristol's designers came up with the Bristol Type 152 Beaufort. Powered by two Bristol Taurus radial piston engines, the Blenheim was armed with four .303 machine guns: two each in nose and dorsal turrets. It could be equipped with either 1,500lb of bombs or mines, or a single 1,605lb torpedo. (Col Dudley Wall)

South Africa received 18 Beaufort Mk Is in 1941, operated by SAAF Nos. 36 and 37 Coastal Flights out of Wingfield, Cape Town, to protect the vital shipping lanes around the Cape of Good Hope. Another 40 were received in June 1943. (russ c)

In 1942, Beauforts flown by SAAF No. 16 Squadron participated in the Madagascar campaign against Vichy French forces. From June 1943, now equipped with ASV radar, the SAAF Beauforts commenced day and night anti-submarine and convoy protection over the Libyan coast. From 1942 to 1943, Beauforts were also operated by SAAF Nos. 20, 22 and 23 Squadrons. (SAAF)

19. Consolidated Catalina

Left: The American Consolidated PBY-5 Catalina flying boat first flew in March 1935 as a long-range naval reconnaissance aircraft. Powered by two Pratt & Whitney R-1830-92 Twin Wasp radial engines, the aircraft had a range of 2,500 miles. (John5199, CC BY-SA 2.0, Wikimedia Commons)

Below: In April 1945, the SAAF received the first of 16 Catalinas. From November 1942, RAF No. 262 Squadron operated land-based PBY-5s from Congella harbour at Durban on South Africa's east coast. During this period, a base was also established at St Lucia on the Natal coast. In the light of a threatened Japanese invasion, the squadron concentrated on anti-submarine patrols. (Col Dudley Wall)

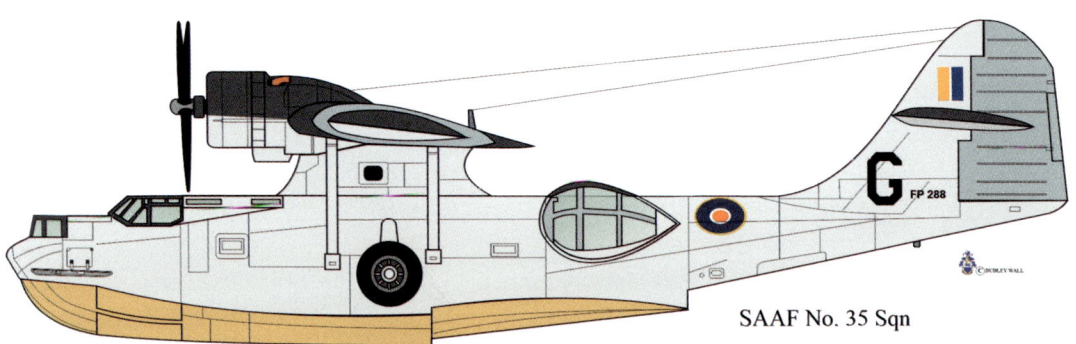

SAAF No. 35 Sqn

Late in 1943, a detachment of seven Catalinas moved to AFB Langebaan in the Cape Province, where conversion training of SAAF crews commenced. By this time, SAAF personnel had already been incorporated into the RAF squadron. In February 1945, Britain disbanded the squadron, which was then renumbered SAAF No. 35 Squadron. The Catalinas remained in service with the SAAF until January 1957. (SAAF)

20. Curtiss Tomahawk

SAAF No. 5 Sqn

Curtiss Tomahawk IIB was the British name given to the Curtiss P-40C, the second version of the P-40 manufactured for the United States Army Air Corps. Armed with six .303 machine guns, the Tomahawk was not suited for combat in Europe as it had been optimised for combat below 15,000 feet. Therefore, in North Africa, where aerial combat tended to be at lower altitudes, the Tomahawk was fairly well matched against Axis Messerschmitt Bf 109 and Fiat G-50 fighters. (Col Dudley Wall)

In September 1941, upon arrival in Egypt from Kenya, SAAF No. 4 Squadron was equipped with Tomahawks. Nos. 5 and 2 Squadrons followed in December (from AFB Swartkop) and April 1941 (from Kenya) respectively, where they too took delivery of Tomahawks. The former was initially tasked with shipping patrols, for which the Tomahawk provided efficient top cover. (SAAF)

On 6 December 1942, the three SAAF Tomahawk squadrons in the Western Desert were assigned to SAAF No. 7 Wing, the first time South African squadrons had operated together as a unified wing. The wing badge was a leaping hartebees on a red shield and was painted on the rudder of wing aircraft. (Articseahorse, Wikimedia Commons)

21. Douglas Boston

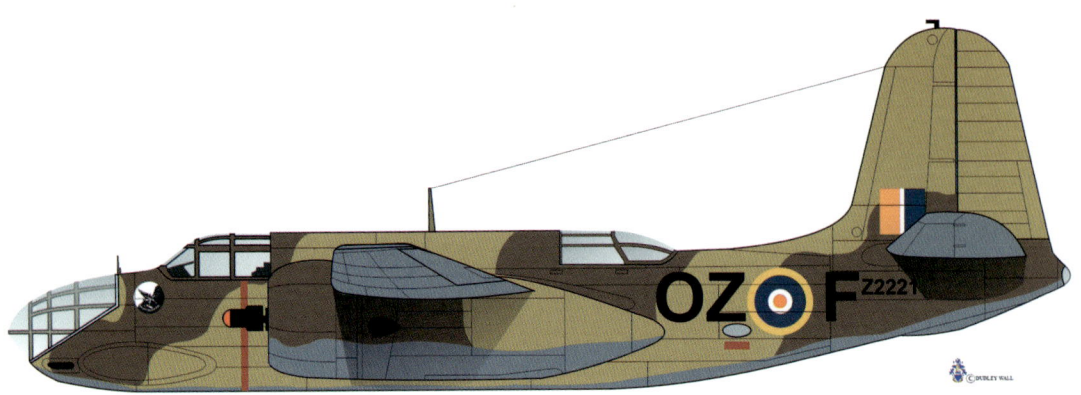

SAAF No. 24 Sqn

The Douglas A-20C (DB-7B) Havoc was an American light bomber, attack aircraft, night intruder, night fighter and reconnaissance aircraft of World War Two. The RAF and Commonwealth forces gave it the service name Boston. It was powered by a pair of Wright R-2600-A5B radial engines. (Col Dudley Wall)

Left: Equipped with Boston Mk IIIs, SAAF Nos. 12 and 24 Squadrons, 3 (SA) Light Bomber Wing, Mediterranean Air Command, served in North Africa in the role of a daytime tactical bomber. Targets for these aircraft included Axis gun positions, lines of communication and troop concentrations. No. 12 Squadron's Bostons saw combat during the Second Battle of El Alamein (23 October to 11 November 1942). At the time, it was said the Germans referred to combined SAAF Nos. 12 and 24 Squadrons' missions as the '18 imperturbables' for their tight attack formations during ever-increasing day and night sorties while contending with heavy enemy flak. By 1943, the SAAF's Bostons had been replaced. (SAAF)

With a crew of four, the Boston was armed with eight .303 machine guns in a fixed, four-gun nose installation and twin-gun dorsal and ventral positions, plus provision for up to 2,000lb of bombs carried internally. (SAAF)

22. Gloster Gladiator

First introduced in 1937, the British Gloster Gladiator biplane fighter was rendered obsolete by newer monoplane designs even as it came off the production line. The single-seat interceptor biplane was armed with two fuselage-mounted and two underwing .303 machine guns. (Mike Freer, GFDL V 1.2, Wikimedia Commons)

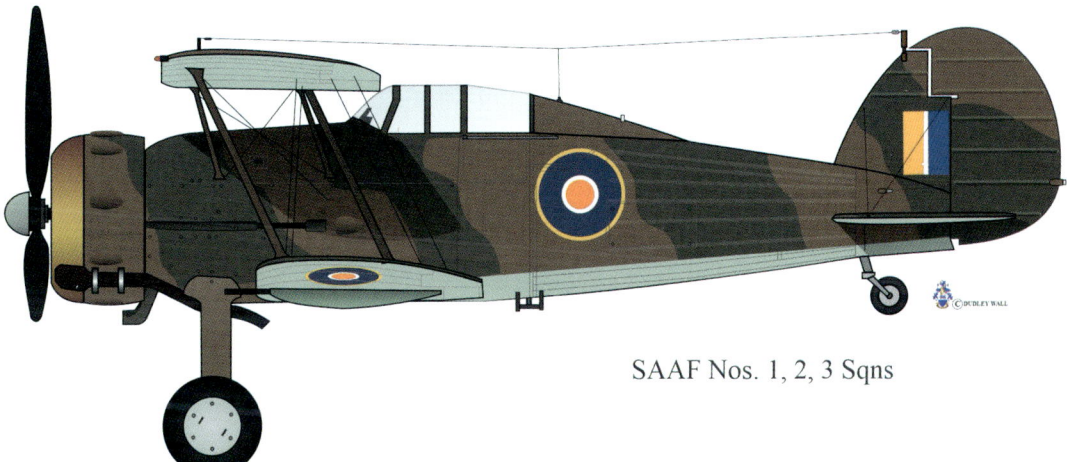

SAAF Nos. 1, 2, 3 Sqns

South Africa received 11 ex-RAF Gladiator Mk IIs for use in the East Africa Campaign as a result of a shortage of Hurricanes. These were supplied to SAAF Nos. 1, 2 and 3 Squadrons. At the end of the campaign, No. 1 Squadron's surviving Gladiators were taken on charge by No. 237 (Rhodesia) Squadron, Southern Rhodesia Air Force. (Col Dudley Wall)

In November 1940, SAAF No. 1 Squadron with eight Glosters was based in Gedaref, south-eastern Sudan. At the start of the British offensive into Ethiopia on the 6th, eight Fiat CR.42 fighters of the Italian Regia Aeronautica's 412a Squadriglia virtually eliminated K Flight and No. 1 Squadron as a fighting force by shooting down six Glosters. Flying Gladiator II N5855, Squadron Leader Major 'Schalk' van Schalkwyk of No. 1 Squadron died of his wounds in captivity the following day. (SAAF)

23. Hawker Hurricane

SAAF No. 40 Sqn

By the outbreak of World War Two, 19 RAF squadrons were fully equipped with single-seat Hurricane Mk I fighter-bombers, the latest and most advanced warplane from the British Hawker aviation stable. Powered by a Rolls-Royce Merlin XX piston engine, the Mk II was armed with 12 .303 forward-firing machine guns (the Mk I had eight), and could carry two 250lb or one 500lb bombs. (Col Dudley Wall)

In February 1939, the SAAF took delivery of its first seven Hurricane Mk Is, which were assembled at Stamford Hill Aerodrome, Durban, before being flown to AFB Waterkloof to equip SAAF No. 3 Squadron. By October 1940, the Hurricanes were in combat in East Africa, where they joined SAAF No. 1 Squadron. (SAAF)

A total of 29 Hurricane Mk Is and 136 Hurricane Mk IIA, B and Cs served in the SAAF. By January 1943, Hurricane-equipped SAAF No. 40 Squadron (Gremlins) was flying in a tactical reconnaissance role, supporting the advance of the British Eighth Army after the Second Battle of El Alamein. Missions were executed deep into Tripolitania, Libya, as the Afrika Korps and their Italian allies retreated into defeat. (asisbiz)

24. Curtiss Kittyhawk

SAAF No. 2 Sqn

The Curtiss P-40 Warhawk was the third most-produced American fighter of World War Two, and when production ceased in November 1944, 13,738 had been built. The British Commonwealth used the name Kittyhawk for models equivalent to the P-40D and all later variants. The Kittyhawk was equipped with six .5in machine guns in the wings and provision for one 500lb bomb under the fuselage. (Col Dudley Wall)

During the critical turning point in the Western Desert campaign in June and July 1942, Hurricanes of SAAF Nos. 2, 4 and 5 Squadrons, No. 233 Wing, escorted SAAF Boston bombers. This was described by most historians as the SAAF's finest hour. In July alone, the SAAF Hurricane fighters flew 2,344 sorties. (SAAF)

In 1942, SAAF No. 5 Squadron received Kittyhawk IIIs and later Kittyhawk IVs and began to specialise in the ground-attack role, although was still retained as an ordinary fighter squadron as required. (SAAF)

25. Curtiss Mohawk

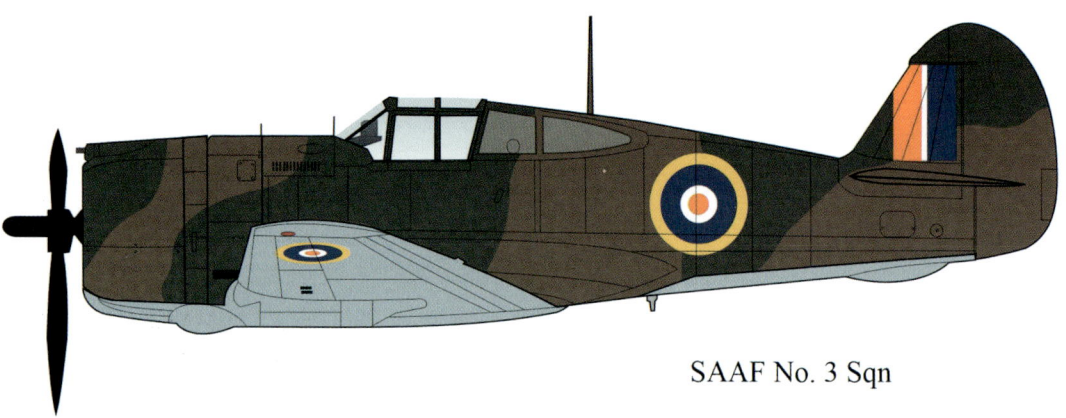

SAAF No. 3 Sqn

The Curtiss P-36 Hawk Model 75 was an American-built fighter of the 1930s and '40s, the first of a new generation of combat aircraft. Within the Commonwealth, the type was usually referred to as the Mohawk. Considered obsolete for the European theatre, 229 Mohawks nonetheless served with Britain, mostly from intercepted supplies for occupied France; 72 of these were sent to the SAAF. (Col Dudley Wall)

On 26 February 1942, SAAF No. 6 Squadron was formed at AFB Swartkop with the Curtiss Mohawk IV. Formed at AFB Waterkloof in September 1940 for service in Kenya, SAAF No. 3 Squadron operated Mohawk IVs out of Alomata airfield, East Africa. The Mohawk was in service with the SAAF from 1941 to 1942. For some time it was strongly believed that a Japanese invasion was imminent anywhere along Africa's Indian Ocean coast. In a proactive move, the SAAF reinforced coastal defences with the deployment of two Mohawk mobile fighter squadrons. Due to the potential threat of Japanese submarines, which did, in fact, venture into South African waters to look for British warships, SAAF No 5 Squadron was deployed to Groutville and No 6 Squadron to Stanger, both on the Natal north coast. Late in 1941, the Mohawk IV was also used in Abyssinia by SAAF No 41 Squadron. (SAAF)

26. Fairey Battle

Developed during the mid-1930s as a monoplane successor to the Hawker Hart and Hind biplanes, the three-seater Battle light bomber was equipped with the same high-performance Rolls-Royce Merlin piston engine that powered various contemporary British fighters such as the Hawker Hurricane and Supermarine Spitfire. Three-seat light bomber. (RAF)

By August 1940, the entire SAAF No. 11 (Bomber) Squadron had converted to Battles, operating in Italian Somaliland and Ethiopia during the East Africa Campaign. In mid-1941, No. 11 Squadron was disbanded and the whole Battle fleet passed on to SAAF No. 15 Squadron. In total, the SAAF received around 340 Battles, including the trainer (T) and target tug (TT) variants. At No. 41 Air School, East London, they were used for gunnery training. Armament included a .303in machine gun in each of the starboard wings and rear cockpit. It had a bombload capacity of 1,000lb. Apart from the Battles used on operational duties, the SAAF received 150 Battles for training purposes, including those serving with SAAF Nos. 41, 42 and 43 Air Schools when South Africa joined the British Commonwealth Air Training Plan. It was usually described as a robust, easy to fly aircraft, even for inexperienced pilots. However, by the time the Battle had entered service, its features had been rendered entirely obsolete by the fast advancement in aircraft technology at the time. (Col Dudley Wall)

SAAF No. 15 Sqn

27. North American Harvard

Acquiring its first North American AT-6 Texan trainer in December 1938, the RAF named it the Harvard. The name stuck, and ultimately Britain and Commonwealth countries took on more than 5,000 Harvards as their premier trainer. Throughout South Africa, the superior Harvard Mk II replaced the Miles Master and Hawker Hart variants at SAAF training schools. By the end of the war, the SAAF had taken delivery of more than 630 Mk I and Mk II Harvards. (Col André Kritzinger, CC BY-SA 3.0, Wikimedia Commons)

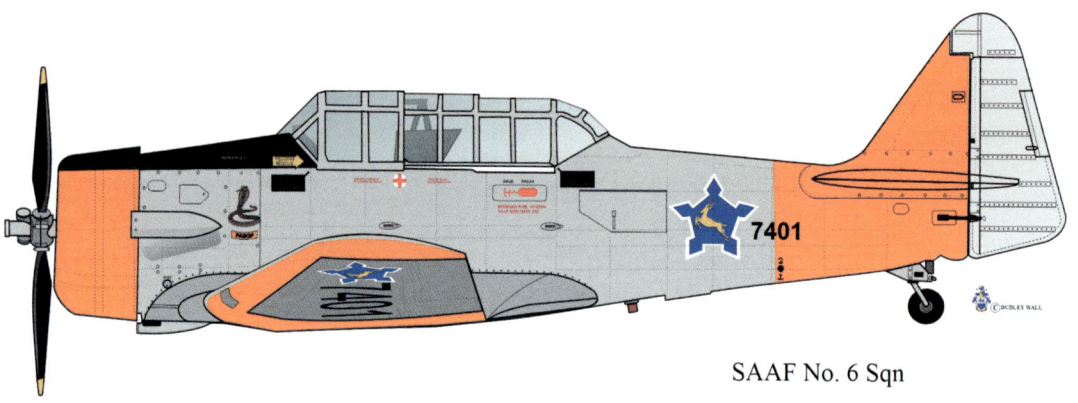

SAAF No. 6 Sqn

Above: Despite the fact that many of the Harvards had to be returned under the Lend-Lease Scheme, the SAAF retained large enough numbers to equip five regular and six Citizen Force squadrons in full or as auxiliary aircraft. (Col Dudley Wall)

Left: From 1952 to 1954, the SAAF acquired 65 refurbished AT-6As and AT-6Cs and 30 rebuilt T-6Gs. The final Harvards to be purchased were four Mk IIAs and Mk IIIs from the Belgian Air Force in 1961. To celebrate 55 years of service upon retirement, in 1995, 55 SAAF Harvards performed a flypast at AFB Langebaanweg. (NJR ZA, Wikimedia Commons)

28. Bristol Blenheim

The Bristol Blenheim was a twin-engine, high-performance medium bomber powered by Bristol Mercury VIII air-cooled radial engines. The Blenheim was armed with a .303 Browning machine gun in the port wing, one or two rear-firing in an under-nose blister, and two in the dorsal turret. It could carry 1,200lb of bombs in internal and external configurations. (Airwolfhound, CC BY-SA 2.0, Wikimedia Commons)

Given the nickname 'Bisley', Blenheim Mk IVs of SAAF Nos. 15 and 17 Squadrons saw active service in North Africa and the Sudan. Upon the end of the East Africa Campaign, SAAF No. 16 Squadron moved to Kenya where it flew Blenheims on maritime patrols.

On 4 May 1942, 12 crewmen of SAAF No. 15 Squadron boarded three Blenheim Mk IVs and took off on a training and familiarisation exercise from the Kufra Oasis in the Libyan desert. In a tragic litany of navigational errors, the trainees became totally lost. Even though they made a successful emergency landing, all but one perished in the severe conditions. (Airwolfhound, CC BY-SA 2.0, Wikimedia Commons)

29. de Havilland Mosquito

Despite being made of wood and with no armament, the de Havilland Mosquito went on to become one of the most successful and popular aircraft of World War Two. Two Rolls-Royce Merlin engines gave the light, low-drag frame the power to reach 300mph. (Alan Wilson, CC BY-SA 4.0, Wikimedia Commons)

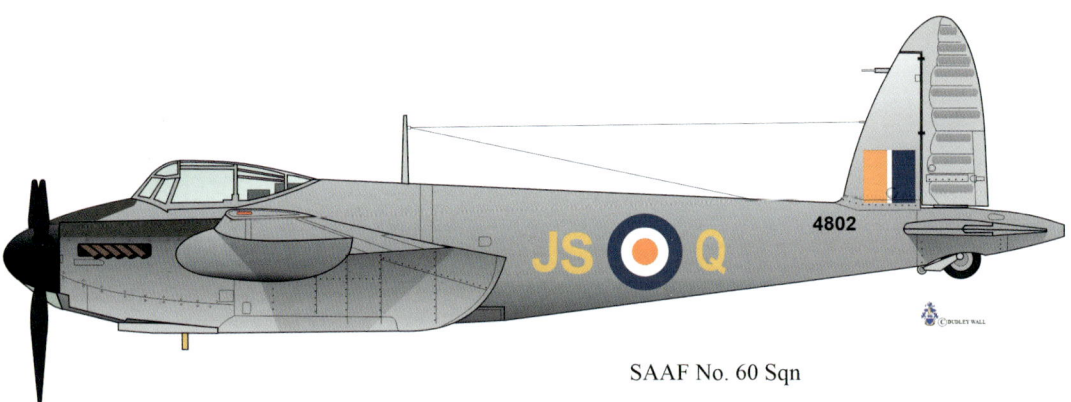

SAAF No. 60 Sqn

In February 1943, SAAF No. 60 Squadron acquired their first photo reconnaissance (PR) Mosquito IIs at the instigation of Field Marshal Bernard Montgomery while serving in North Africa. The squadron was subsequently sent to US AFB San Severo in Italy, and in February 1944 was equipped with Mosquito PR XVIs in preparation for the southern European campaign. They carried out large-scale photographic surveys of parts of Sicily and other Axis-held areas, eventually ranging over the Alps and deep into Austria and Germany, where, towards the end of the war, they met with dangerous interceptions by Messerschmitt Me 262 jets. (Col Dudley Wall)

During PR flights in 1944, SAAF No. 60 Squadron Mosquito PR Mk XIs conducted several photographic sorties over the Auschwitz-Birkenau camps, in which the Holocaust was visible but was not recognised. (SAAF)

30. Douglas DC-3/C-47 Dakota

Above left: During World War Two, the SAAF received 84 Lend-Lease Dakotas from the RAF. In Egypt, two newly formed SAAF squadrons – Nos. 28 and 44 – received the C-47A Mk III and C-47B Mk IV transports. SAAF No. 44 Squadron ended up at Bari in Italy, air dropping war matériel and supplies to Yugoslavian partisans. At the end of the war, the SAAF's Dakotas ferried thousands of South African troops back home. (SAVA)

Above right: Known by the Americans as the 'Gooney Bird' and the 'C-47 Skytrain', the World War Two workhorse was known by the rest of the English-speaking world simply as the 'Dak'. And to most troops who travelled in the transport, it had the uncomplimentary but affectionate moniker of 'Vomit Comet'. The Dakota is capable of transporting 10,000lb of cargo or 27 persons. (JMK, GFDL V 1.2, Wikimedia Commons)

Dakotas have been operated by SAAF No. 25 (AFB Ysterplaat), No. 27 (AFB Ysterplaat), No. 28 (AFB Waterkloof), No 44 (AFB Swartkop), No. 60 (AFB Waterkloof) Squadrons and No. 86 Advanced Flying School (AFB Bloemspruit). (Herman Potgieter)

During the Border War, a specially kitted out Dakota, nicknamed 'Spook', was employed solely as an aerial communications relay platform, flying at 16–17,000 feet. Some Dakotas were converted into gunships called 'Dragons', fitted amidships with a 20mm cannon. Included in these adaptations was a 'sky shout' facility consisting of large loudspeakers to disseminate propaganda to the enemy below. (Bob Adams, CC BY-SA 2.0, Wikimedia Commons)

31. Martin Baltimore

The Martin 187 Baltimore was an American four-seat light bomber with a bomb load of up to 2,000lb. The aircraft was heavily armed, with four .303 wing-mounted machine guns, a Vickers K machine gun in the dorsal turret and two machine guns in a ventral position, with provision for similar guns in fixed rear-firing positions. The Commonwealth designation was A-30. (RAF)

During World War Two, the SAAF took delivery of around 125 Baltimores: No. 15 Squadron operated Baltimore IIIs and Vs in the Mediterranean (1943–45); No. 21 Squadron Baltimore III and IVs in North Africa and Italy (1942–44); and No. 60 Squadron Baltimore II and IIIs in North Africa (1942–43) Only two Baltimore Mk. Is served in the SAAF. Generally popular, the Baltimore was nonetheless hampered by its narrow fuselage; movement around the aircraft by the four-man crew was largely impossible, which made for cramped missions. In July and August 1942, the Desert Air Force altered its tactics so that the Baltimores flew at higher altitudes to place them beyond the range of light flak units as well as fighter-bomber escorts. (SAAF)

32. Bristol Beaufighter

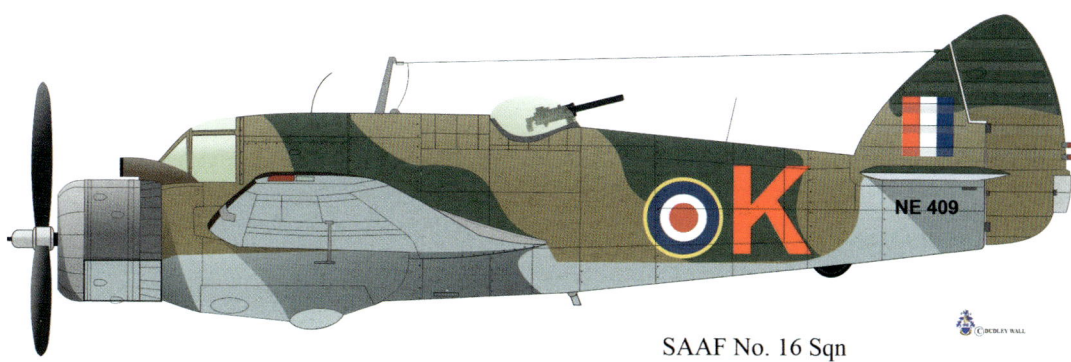

SAAF No. 16 Sqn

The Bristol Type 156 Beaufighter was a British multi-role aircraft developed during World War Two. The 'Beau' was a formidable attack/strike platform, packing an arsenal of four forward-firing 20mm cannon and six .303 machine guns, plus a Vickers K gun in the dorsal turret. In addition, it could carry an external 18in Mk XII torpedo, while underwing hardpoints accommodated eight 60lb RP-3 rockets. During World War Two, the Beaufighter also played a significant role in the Battle of Britain, protecting the skies over the south of England. Flying at night, all-black painted Beaufighters acted as night interceptors, their large size allowing the fitment of heavy armament and early airborne interception radar without major performance drawbacks. (Col Dudley Wall)

Equipped with Beaufighters in November 1943, Egypt-based SAAF No. 16 Squadron flew anti-shipping missions in the Aegean before moving to Italy to join the Balkan Air Force in support of Yugoslavian partisan operations. Wartime SAAF No. 19 Squadron (renumbered RAF No. 227 Squadron) flew Beaufighter VIs and Xs to great effect on bombing and strafing attacks on German infrastructure in Greece and Yugoslavia. After also joining the Balkan Air Force, this squadron employed the newly acquired 60lb rockets on its Beaufighters to devastating effect. (RAF)

33. Consolidated B-24 Liberator

With 18,500 built, the American Consolidated B-24 Liberator is the world's most produced multi-engine heavy bomber. Powered by four Pratt & Whitney R-1830-35 Twin Wasp, R-1830-41 or R-1830-65 14-cylinder, two-row air-cooled turbo-supercharged radial piston engines, depending on range, the Liberator could carry between 2,700lb and 8,000lb of bombs. It was armed with ten .5in M2 Browning machine guns in four turrets and two waist positions and had a crew of 11. (RAF)

SAAF Nos. 31 and 34 Squadrons were formed in April 1944 under SAAF No. 2 Wing based at Foggia, Italy, where they saw active service flying B-24J Liberators during World War Two. The two squadrons are most famous for flying 196 missions from Italy to Warsaw with supplies (12 330lb canisters in each Liberator) during the uprising of the Polish resistance. On the weekend of 13–16 August, 25 SAAF Liberators were shot down and 69 South African airmen were killed. Later on in the war, the SAAF squadrons also dropped supplies to Marshal Tito's Yugoslavian resistance fighters. Most resupply drops were carried out at night, which made it easier for the aircrews to locate the partisans on the ground. It was dangerous for both them and the aircrews, as the partisans risked capture by the Germans and many were shot when they were discovered. The aircraft had to fly very low, always with the risk of crashing into the side of a hill or mountain. Flying Liberator Mk IVs, SAAF No. 31 Squadron operated in northern Italy, the Balkans, Austria and Southern France, as well as performing mine-laying operations along the Danube. Daylight missions were even more dangerous for both the aircrews and the partisans on the ground, as the low-flying bombers could easily be seen on their low-altitude slow approach to the drop zones. With several aircraft in formation on their final run over the drop zone, they were an easy target for German troops on the ground and Luftwaffe fighter patrols. (Col Dudley Wall)

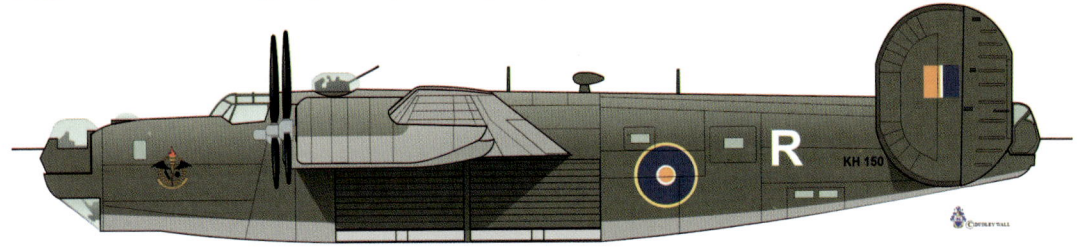

SAAF No. 34 Sqn

34. Martin Marauder

The first American B-26 Marauder twin-engined medium bomber flew on 25 November 1940. The aircraft was armed with 11 .5in machine guns: one trainable nose position, four fixed in blisters on the fuselage and operated from the cockpit, two in the dorsal turret, two in the tail turret, and one each in port and starboard lower waist positions. It could accommodate up to 4,000lb of bombs. (RAF)

A total of 100 Marauder B-26 Mk IIs served in the SAAF. The four SAAF Squadrons at Pescara and Iesi, Italy – Nos. 12, 21, 24 and 30 – formed the all-Marauder No. 3 (SA) Wing, Desert Air Force, while SAAF No. 25 Squadron flew Marauders on Yugoslavia missions. In 1943, deliveries of 100 long-wingspan B-26C-30s (Marauder II) allowed two squadrons of the SAAF – Nos. 12 and 24 – to be equipped, these being used for bombing missions over the Aegean Sea, Crete and Italy. A further 350 B-26Fs and Gs were supplied in 1944, with two more SAAF squadrons (Nos. 21 and 30) joining Nos. 12 and 24 in Italy to form an all-Marauder-equipped wing, while one further SAAF squadron (No. 25) and a new RAF squadron (No. 39), re-equipped with Marauders as part of the Balkan Air Force. Several Marauders from No 3 Wing were modified, immediately after the war ended, for passenger and freight carriage. (Col Dudley Wall)

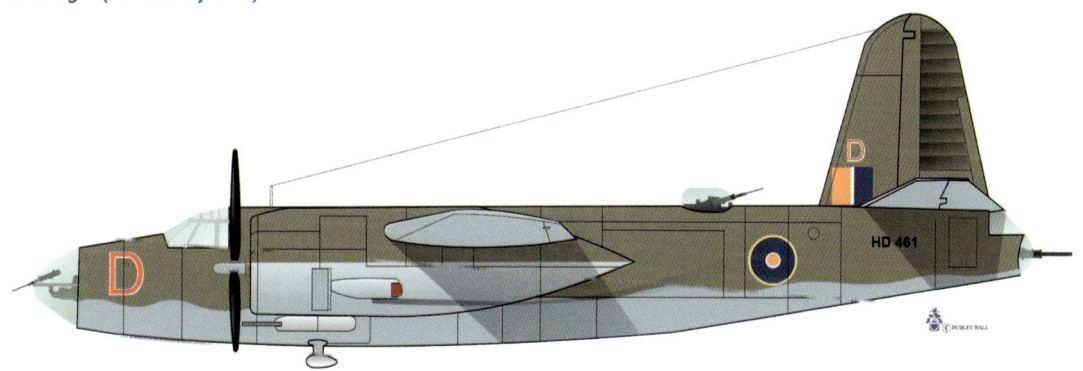

SAAF No. 30 Sqn

35. Supermarine Spitfire

The iconic Supermarine Spitfire, famed for saving mainland Britain from the Luftwaffe, was produced in 24 marks and numerous sub-variants within those marks. The Mk V was one of the most successful 'stop-gaps' ever introduced into RAF service, with the VBs becoming the main production variant. This would be the first pressurised Spitfire. More than 140 RAF squadrons and nine countries, including South Africa, operated the type. (Col Dudley Wall)

The SAAF acquired Mk VB and Mk VC types, including the tropical version, the former with characteristic clipped wingtips (LF) to enhance its low-altitude performance, most notably its roll speed. The Mk VB was armed with two 20mm cannon and four 3.03 machine guns in the wings. SAAF Nos. 1, 3, 7, 10, 11, 40 and 41 Squadrons were equipped with Mk Vs, operating in the North Africa, Middle East, Mediterranean and Italian theatres. (Bob Adams, CC BY-SA 2.0, Wikimedia Commons)

Following the end of World War Two, Britain offered its Commonwealth air forces a gift of surplus Spitfires in recognition of their role during the war. In 1947–48, the SAAF received 136, including 80 gifted Spitfire Mk IXs, while a further 56 Mk IXs were purchased at £2,000 each. They were mainly used for training. (SAAF)

36. Vickers Wellington

SAAF No. 26 Sqn

The Vickers Wellington was a British twin-engined, long-range medium bomber designed in the mid-1930s. With a crew of six, the Wellington was armed with six to eight .303 Browning machine guns and could carry up to 4,500lb of bombs. The most mass-produced British bomber of World War Two, the Wellington was designed by Vickers-Armstrong's Chief Designer, Rex Pierson, utilising the geodetic construction methods devised by Barnes Wallace and used in the earlier Vickers Wellesley. The prototype was flown at Brooklands on 15 June 1936, and production examples served with distinction throughout World War Two, despite eventually being superseded in their primary role by the much larger 'heavy bombers' such as the Avro Lancaster. The "Wellie" was to form the spearhead of Bomber Command's offensive against Germany, making up some 60 per cent of the numbers in the first 1,000-bomber raid on 30 May 1942. (Col Dudley Wall)

The SAAF operated Wellington B.Mk VIIIs (Type 429) and B.Mk XIs (Type 458) in a reconnaissance/anti-submarine role. SAAF No. 26 Squadron employed Wellingtons on reconnaissance and anti-submarine patrols from Takoradi, Gold Coast (now Ghana), from 1943 until the end of the war. Late in 1944, SAAF No. 17 Squadron commenced Wellington conversion training in Egypt. (RAF)

37. de Havilland Dove/Devon

The de Havilland DH.104 Dove was a short-haul, small airliner developed in Britain in the 1940s. Unlike its predecessor, the DH.89 Dragon Rapide biplane, the twin-engine Dove was an entirely metal construction. It was operated by South African Airways, but subsequently reconfigured as a military communications variant, designated the Devon C1, employed by the RAF and the Royal Navy as a light transport and command platform. (SAAF)

In September 1945, SAAF No. 28 Squadron returned permanently to South Africa and was based at AFB Swartkop where, in 1949, it took delivery of nine Devons. These were added to the SAAF's VIP fleet. The aircraft were retired in 1964. (Bob Adams, CC BY-SA 2.0, Wikimedia Commons)

38. Douglas DC-4

At the end of World War Two, the American Douglas company resumed production of the DC-4-1009, the commercial variant of the wartime C-54. From 1945 to 1947, South African Airways acquired six DC-4-1009 airliners and an ex-USAAF C-54, which operated the 'Springbok Service' from Johannesburg to London until 1950, when they were replaced with Lockheed Constellations. (SAAF)

In 1966, the five remaining DC-4-1009s were transferred to the SAAF to be flown by SAAF No. 44 Squadron out of AFB Swartkop. The DC-4 was widely used as a passenger and VIP transport aircraft. A far lesser known role of the DC-4 was that of electronic warfare. This involved electronic reconnaissance and real-time support of aerial strikes during the Border War. (Aeroprints, CC BY-SA 3.0, Wikimedia Commons)

39. Sikorsky S-51 Dragonfly

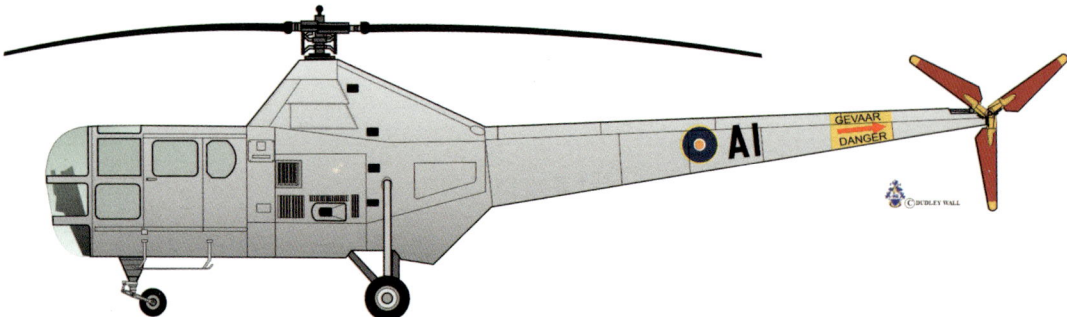

The American Sikorsky/Westland H-5 (civil designation S-51) utility helicopter went into licensed production in Britain with Westland in December 1946. During World War Two, several variants of the Sikorsky R-4 and R-6 were used by the American and British militaries. Sikorsky intended to develop a civil variant of its XR-5, the original designation of the S-51, for the post-war market. The Westland WS-51 Dragonfly was not, however, an exact copy of the American model. It was fitted with a more powerful Alvis Leonides 500 engine in place of the American Pratt & Whitney. The prototype Dragonfly made its first test flight on 5 October 1948 and the first European operator to employ this helicopter was British European Airways (BEA) along with Pest Control Ltd, whose helicopters were equipped with a purposely designed pressure-spray system. Despite the fact that it was a good helicopter, this model did not enjoy commercial success in the civil field. (Col Dudley Wall)

The S-51 was the first helicopter to enter SAAF service when three aircraft (A1, A2 and A3) were acquired in 1948, and allocated to the re-formed SAAF No. 12 Squadron for the purpose of spraying the tsetse fly in Zululand. (Alan Wilson, CC BY-SA 4.0, Wikimedia Commons)

40. Short Sunderland

The Short S.25 Sunderland was a British flying boat patrol bomber developed for the RAF. It was one of the most powerful and widely used flying boats throughout World War Two. The Short Sunderland S.25, designed and built by Belfast-based Short Bros, remained in front-line service for over 20 years. It was also the last flying boat operated by the Royal Air Force. The Sunderland was produced as a military development of the S.23 'C'-Class or Empire flying boat operated by Imperial Airways. It entered service in June 1938 and was the first British flying boat to have power-operated gun turrets as part of its defensive armament. This strong protective armament resulted in the Germans giving it the nickname 'Flying Porcupine'. The Sunderland was roomy enough to give the crew of ten or more men some comfort on their long patrol flights, which could last up to 13 hours. The front part of the fuselage was divided into two decks. The upper deck contained the cockpit with the two pilots, plus the stations for the flight engineer, the wireless operator and the navigator. On the lower deck there was a bomb room, where bombs or depth charges were stored on movable racks that were moved to under the wing before an attack. (RAF)

Following the end of World War Two, Britain offered its Commonwealth air forces a gift of surplus aircraft in recognition of their role during the war which, among others, included 12 Sunderland GR Mk Vs to the SAAF for service in a maritime patrol and transport role. A further three were purchased to maintain the 15-aircraft fleet already with No. 35 Squadron. (Col Dudley Wall)

SAAF No. 35 Sqn

41. de Havilland Heron

First flown in 1950, the DH.114 Heron airliner was effectively a stretched four-engine version of the DH.104 Dove/Devon. Unpressurised and flown by a crew of two, it was able to carry up to 17 passengers. It was powered by four de Havilland Gipsy Queen 30-2 engines. Dove outer wing panels were used on the Heron, and Dove nose and tail units were joined by an extended fuselage. As with the Dove, the airframe, engines and propellers were all made by de Havilland. It was intended for 'outback' operations into airfields with minimal facilities. (Mike Freer, GFDL V 1.2, Wikimedia Commons)

Operated by SAAF No. 28 Squadron out of AFB Waterkloof, two Heron 2Bs transports (Nos. 120 and 121) went into service with the SAAF in 1955 and were retired in 1962. (kitmasterbloke, CC BY-SA 2.0, Wikimedia Commons)

42. de Havilland Vampire

Narrowly missing World War Two active service, in 1950 the twin-boomed de Havilland Vampire fighter replaced the SAAF's aging Spitfire Mk IXs, catapulting the force into the jet age. SAAF No. 1 Squadron at AFB Waterkloof took delivery of ten FB.5s. Some of these were operated by the squadron at AFB Langebaan in a training capacity. Up to 1956, a further 67 Vampires were acquired, including the FB.6, FB.52, and T.11 and T.55 training variants. (Alan Wilson, CC BY-SA 4.0, Wikimedia Commons)

The Vampire's armaments included four 20mm Hispano cannon in the nose, and external ordnance configurations of eight 60lb rockets and two 500lb bombs, or two 1,000lb bombs. (Alan Wilson, CC BY-SA 4.0, Wikimedia Commons)

From 1956, Canadair Sabres were introduced to the squadron, resulting in most of the Vampires being transferred to the operational school at Langebaanweg. By the end of the 1960s these found a new home at Pietersburg, where for a short period SAAF No. 1 Squadron operated FB.52s and T.55s. The introduction of the Atlas Impala saw all but two of the Vampires withdrawn by the end of 1972. (Alan Wilson, CC BY-SA 4.0, Wikimedia Commons)

43. Vickers Viscount

This British medium-range turboprop airliner was first flown in 1948. In June 1958, the SAAF took delivery of its only Vickers Viscount, named *Casteel*. A 781D variant, capable of seating 20 passengers, *Casteel* joined SAAF No. 21 Squadron's VIP flight. (Matthew Janse van Rensburg)

The most successful of Britain's post-World War Two airliners, the Vickers-Armstrong Viscount had its origins as the Brabazon Committee's Type IIB. When introduced, it was a revelation on British European Airways' (BEA) European and domestic routes, compared with the Douglas DC-3s that it replaced. The Type 700 series first flew on 19 April 1950 and its smoothness, good operating economics and pressurisation contributed to its success including, surprisingly, in North America. The SAAF's only Viscount, *Casteel* (c/n 280), was powered by four Rolls-Royce Dart 510 turboprop engines, while some of the original production Series 700 Viscounts were acquired by Central African Airways and South Africa purchased seven of the later Series 813 aircraft and, much later, two Series 818 from Fidel Castro's Cuba. In November 1991, *Casteel* was sold to Field Aviation (South Africa) Ltd, and early in 1996, it was purchased by Zaire-registered Bazair. On 6 June 1997, the aircraft crashed in the Democratic Republic of the Congo, killing all 23 crew and passengers on board. (SAAF)

44. de Havilland Canada Chipmunk

The Canadian DHC.1 Chipmunk was essentially a replacement for the Tiger Moth biplane primary trainer. The aircraft featured low-mounted wings and a two-seat tandem cockpit, which was fitted with a clear Perspex canopy to provide all-round pilot/student visibility. (Gerry van Tonder)

Whilst the SAAF never flew the Chipmunk, a crashed Royal Rhodesian Air Force Mk 10 (WG354) was acquired by the SAAF Museum in 1979. The aircraft was restored by SAAF No. 35 Squadron and placed with the museum's historical flight. (Bob Adams, CC BY-SA 2.0, Wikimedia Commons)

45. North American Mustang

Arising out of a requirement for high-performance fighters, and following hitherto poor operational performance, the North American F-51 Mustang underwent a major redesign, centred on replacing the Allison engine with the Packard-built Rolls-Royce Merlin. Entering service with the US Air Force (USAF) in December 1943 escorting Allied bombers, the long-range, high-altitude Mustang proved to be the most effective fighter over Europe in the last two years of World War Two. (USAF, Public Domain)

SAAF No. 2 Sqn

The Mustang was armed with six .5in machine guns, plus up to two 1,000lb bombs or six 5in rocket projectiles and napalm. (Col Dudley Wall)

Above left: In September 1950, volunteer pilots of SAAF No. 2 'Flying Cheetahs' Squadron entered the Korean War theatre, where they were equipped with USAF-owned F-51D Mustangs. The SAAF already had experience with the P-51 Mustang when serving with SAAF No. 5 Squadron during World War Two. Although flying in SAAF colours, these aircraft were, however, never placed on the SAAF inventory. (USAF, Public Domain)

Above right: The SAAF flew with the distinctive Springbok in the centre of the roundel, introduced when SAAF No. 2 Squadron went to Korea. Their role was close air support against enemy positions to soften them up for ground attacks, interdiction against communist logistic and communication lines, providing protective cover for rescue operations, reconnaissance flights, and to a lesser extent, interception of enemy aircraft. During the southward advance of Chinese forces into Korea, these pilots attacked enemy troops, trucks and supplies daily in near-zero temperatures. (USAF, Public Domain)

46. Avro Shackleton

The SAAF was equipped with the M3 version of the Avro 696 Shackleton, powered by four Rolls-Royce Griffon 57A piston engines. (Col André Kritzinger, CC BY-SA 3.0, Wikimedia Commons)

This ocean patrol aircraft was equipped with two removeable 20mm Hispano cannons in the nose. The cavernous bomb bay could be fitted with an array of ordnance depending on mission needs, including three Mk 30 or Mk 40 torpedoes or depth charges, or nine 250lb bombs or nine Sonobuoys for underwater detection. The underwings had hardpoints for rocket rails and, when fulfilling an air-and-sea rescue role, a Saro airborne lifeboat could be affixed below the bomb bay. (Col André Kritzinger, CC BY-SA 3.0, Wikimedia Commons)

Operated by SAAF No. 35 Squadron out of D.F. Malan Airport (now Cape Town International), the SAAF took delivery of the first of eight Shackletons in May 1975. After 27 years of exemplary service, the last of the Shackletons were retired late in 1984. During this time the aircraft had been rebuilt, with new wing spars to extend serviceable life. (SAVA)

47. Canadair/North American Sabre

Above left: The appearance of the MiG-15 in the Korean conflict accelerated the USAF's need to upgrade its fighter/interceptor capabilities, resulting in the deployment of the US 4th Fighter Wing, fully equipped with swept-wing North American F-86 Sabres. Besides missiles, its armament included .5in machine guns or 20mm cannon in the fuselage and rockets or bombs under the wings. (SAVA)

Above right: The SAAF's use of the Sabre started during the Korean War when SAAF No. 2 Squadron had its Mustangs replaced with North American F-86F Sabres on loan from the USAF. Attached to the US 18th Fighter-Bomber Wing, the first Sabres were delivered to the squadron in January 1953, and the last returned to the USAF in October of that year. (Richard Keener, USAF, Public Domain)

The distinguished service of SAAF pilots flying the Sabre in Korea was a contributing factor to South Africa's decision to replace its de Havilland Vampires with this jet fighter. In August 1956, the SAAF took delivery of the first of 34 Canadair CL-13B Sabre 6s, which were initially assigned to SAAF Nos. 1 and 2 Squadrons out of AFB Waterkloof. In 1963, No. 1 Squadron became the home of all the SAAF's Sabres. (SAVA)

In 1967, SAAF No. 1 Squadron moved to Pietersburg where, for several years, the Sabres were South Africa's main first-line fighters. With the arrival of the Dassault Mirage F1, in 1975 all the Sabres were transferred to No. 85 Advanced Flying School. Four years later, the Sabre's long and illustrious service with the SAAF ended when the fleet was grounded for the final time. (Alan Wilson, CC BY-SA 4.0, Wikimedia Commons)

48. English Electric Canberra

In September 1963, the SAAF started to take delivery of six B(I).12 and three T.4 trainer Canberras from Britain. The high-altitude bomber and photo-reconnaissance jets went into service with No. 12 Squadron at AFB Waterkloof. The aircraft's weapons fit included two 1,000lb underwing bombs and nine 500lb bombs in the bomb bay. The reconnaissance fit consisted of up to five Zeiss F-96 cameras contained in an adapted gun-pack 'canoe'. (SAVA)

From 1975 (Operation *Savannah*) to 1979 (Cahama, Southern Angola), the SAAF's Canberras were used extensively on interdiction sorties into Angola and Zambia. In March 1979, Canberra No. 452 was shot down over a target in Angola, killing both crew members on board. (SAVA)

49. Sikorsky S-55

The Sikorsky H-19 Chickasaw, also known by its Sikorsky model number, S-55, was a multi-purpose helicopter used by the United States Army and United States Air Force. It was also licence-built in the United Kingdom by Westland Aircraft as the Westland Whirlwind. (SAVA)

In 1954, the South African Defence Force General Staff budgeted for three S-55s to replace motorboats for inshore search-and-rescue duties. The ten-seater helicopters (Nos. A4 to A6) arrived in 1956–57. These were assembled at AFB Ysterplaat, and became part of the Helicopter Flight at Langebaanweg. At this time, SAAF No. 17 Squadron was re-formed to fly the S-55s. (SAAF)

Following the news in 1958 that the US was phasing out the piston-engine S-55, South Africa's plans to acquire more were shelved in favour of the more versatile turbo-engine Aérospatiale Alouettes. All three S-55s were sold to commercial concerns in 1967, with A4 becoming ZS-HCO, A5 becoming ZS-HCL and A6 becoming ZS-HCM. (Col André Kritzinger, CC BY-SA 3.0, Wikimedia Commons)

50. Sud Aviation Alouette II

In 1960–61, the SAAF acquired eight French-made Sud Aviation (later Aérospatiale) Alouette II SE.3130 light, multi-purpose helicopters. A simple, relatively inexpensive, robust aircraft, upon its release it took over several world records for the range and altitude of helicopters. This was made possible by the use of a new gas turbine engine that allowed high performance at a low engine weight. (SAVA)

From July 1961, SAAF No. 17 Squadron, having moved to AFB Ysterplaat, took delivery of the first Alouette IIs. The extremely versatile aircraft could be employed in medical evacuation, reconnaissance, rescue, light transport, aerial photography and training. In military service, the Alouette II operated in a close-support and counter-insurgency role. (Olga Ernst, CC BY-SA 4.0, Wikimedia Commons)

51. Sud Aviation/Aérospatiale Alouette III

The SAAF's first Alouette III SE.316A helicopters (the 316B came later) arrived at AFB Waterkloof in February 1962. When sufficient numbers had been received, SAAF No. 17 Squadron was split into three: A Flight at AFB Swartkop, B Flight at Bloemspruit in the Orange Free State, and C Flight, which remained at AFB Ysterplaat. (SAVA)

In addition to land-and-sea rescue missions, the aircraft were employed on coastal patrols along the South West African Skeleton Coast. The Alouettes proved to be an indispensable adjunct to national law enforcement, such as in March 1972 when an SAAF No. 16 Alouette III performed 110 sorties in one day during a police cannabis raid in Swaziland's Lebombo Mountains. (Bob Adams, CC BY-SA 2.0, Wikimedia Commons)

Arguably, the Alouette III's greatest worth to South African forces was as a border counter-insurgency asset. Its rapid troop deployment, ground-attack and combat medical evacuation capabilities in the hands of highly experienced pilots became legend. During the Border War, the Alouette provided an effective air-support platform, equipped with various configurations of 7.62mm machine guns or a 20mm Hispano cannon. (Alan Wilson, CC BY-SA 4.0, Wikimedia Commons)

52. Cessna 185

The SAAF received its first 24 Cessna 185A Skywagons, designated U-17 under the American military assistance programme, in mid-1962. Acquired to replace the aging Auster AOP-6s, most of the new Cessnas were allocated to SAAF No. 42 Squadron, based at the School of Artillery at Potchefstroom. At the time, army pilots flew the Cessnas in the aircraft's continued role of air reconnaissance. (SAVA)

Essentially a utility aircraft for the civil market, the robust Cessna proved popular in the hot and dusty environs of southern Africa. In the second half of the 1960s, Cessna strength with the SAAF was 34. The 185s were deployed mainly on photo reconnaissance to assist the South African Police in South West Africa and Rhodesia (now Zimbabwe). (Bob Adams, CC BY-SA 2.0, Wikimedia Commons)

53. Dassault Mirage III

Following the acquisition of the Dassault Mirage IIICZ single-seat interceptor jet in April 1963, over the next ten years the SAAF took delivery of six other variants of this Mach 2 French jet: BZ two-seat trainer, EZ fighter/bomber version of the CZ, RZ and R2Z tactical reconnaissance, and DZ and D2Z advanced conversion trainers. (SAVA)

The Mirage III CZ was armed with two internally mounted 30mm Defa cannon, and equipped to accommodate a range of bombs or missiles on underwing and belly hardpoints. Its nose carries a Thomson-CSF Cyrano I (Ibis) fire-control and navigation system. (SAVA)

Flying Canadair Sabres out of AFB Waterkloof at the time, Korean War-blooded SAAF No. 2 'Flying Cheetahs' Squadron received the Mirage III CZ, BZ, RZ and R2Z variants. In 1966, SAAF No. 3 Squadron was re-formed at AFB Waterkloof as a 'paper squadron', with Mirage III EZs from No. 2 Squadron at its core. In 1975, when No. 3 Squadron received its first Mirage F1s, its Mirage III EZs and DZ trainers were transferred to No. 85 Advanced Flying School at Pietersburg. (SAVA)

54. Lockheed C-130 Hercules

Since its prototype maiden flight in August 1954 at Lockheed's plant in Burbank, California, the C-130 Hercules tactical transport is the most widely used of its type in the world. Designated the C-130A, it spawned a next major variant as the C-130B, which featured more powerful Allison T56-A-7 turboprops driving four-blade propellers. Integral fuel capacity in the wings and a strengthened undercarriage made the Hercules an attractive option for the SAAF to augment its Dakotas. (Kai Hansen)

Above left: Of the 250 C-130Bs built, only 29 were for export, of which SAAF No. 28 Squadron, AFB Waterkloof, received seven in January 1963. During the early 1970s, the squadron's C-130 engines were upgraded from the Allison T56-A-7 to the T56-A-15. (Bob Adams, CC BY-SA 2.0, Wikimedia Commons)

Above right: Missions supported by the SAAF's C-130 fleet since acquisition include humanitarian, peacekeeping, transport, supply drop, search-and-rescue, anti-piracy and diplomatic missions. However, it has become increasingly costly to support the C-130 fleet, as certain spare parts have gone out of production and are no longer readily available. At some point in the near future, this might mean the aircraft will have to be retired from service if no additional funding is made available. (SAVA)

The fleet saw extensive use during the long-running conflict in South West Africa and Angola, supporting South African Defence Force units during the so-called Border War, which ran from 1966 to 1989. (Col André Kritzinger, CC BY-SA 3.0, Wikimedia Commons)

55. Westland Wasp

The British Westland Wasp was specifically developed for the Royal Navy for shipboard use as an anti-submarine and communications platform on frigates and destroyers. The Wasp has a crew of two, with provision for three passengers or a stretcher at the rear of the cabin. (SAVA)

The SAAF started to operate the first of 11 Wasps from late 1963. Although the South African navy does not have a fleet air arm, several vessels were refitted to accommodate Wasps. Recreated at AFB Ysterplaat in 1964 as SAAF 22 Flight and equipped with the newly delivered Wasp helicopters, the unit became a full squadron in May 1976. After 27 years' service, the Wasp was retired in 1990. (Ad Meskens, CC BY-SA 4.0, Wikimedia Commons)

56. Hawker Siddeley Buccaneer

Appropriately calling themselves 'the Pirates', SAAF No. 24 Squadron was re-formed at Lossiemouth in Scotland in 1965, for type-training on the British-made Buccaneer S.50. Taking delivery of 16 two-seater, low-level strike Buccaneers, the squadron would be based at AFB Waterkloof, near Pretoria. Although the SAAF's Buccaneers came with standard carrier folding wings, these were later fixed, as the aircraft were land-based. (Col André Kritzinger, CC BY-SA 3.0, Wikimedia Commons)

Ordered at a time when the West placed great emphasis on the strategic importance of the sea route around the Cape, the Buccaneer could accommodate four 1,000lb bombs, four underwing Mantra rocket pods or a Nord AS30 air-to-surface missile. (Alan Wilson, CC BY-SA 4.0, Wikimedia Commons)

57. Sud Aviation/Aérospatiale Super Frelon

Designed with technical expertise from Sikorsky, the hull shape of the Super Frelon SA.321L attests to its amphibious capabilities. Classed as a medium-range tactical helicopter, the aircraft could transport 27 fully equipped troops. (Col Dudley Wall)

The transport of often excessive mixed loads during the Border War earned the Super Frelon the perhaps uncomplimentary moniker of 'Putco Bus', a direct reference to one of South Africa's largest public transport companies. (SAVA)

58. Aermacchi/Atlas Impala

Right: The acquisition and subsequent manufacture under licence of the Italian Aermacchi MB-326 proved to be a successful decision in the SAAF's Cold War modernisation programme. (SAVA)

Below: The MB-326M was produced in two models for the SAAF: 40 unarmed Italian-built aircraft and around 125 assembled or locally built in the Transvaal by the Atlas Aircraft Corporation from 1966. The latter, known locally as the Impala Mk I, had provision for externally mounted armaments. (SAVA)

Above left: In 1974, Aermacchi delivered seven MB-326Ks to the SAAF and in 1975 another 15 more single-seaters. After that, Atlas commenced local manufacture of this version as the Impala II, fitting it with the Rolls-Royce Viper 540 engine. After initial pilot training on Harvards, pupil pilots went on to the Flying Training School at AFB Langebaan to enhance their skills on the Impala. Operational training was conducted on Mk IIs at No. 85 Advanced Flying School near Pietersburg (now Polokwane). (Bob Adams, CC BY-SA 2.0, Wikimedia Commons)

Above right: With the commissioning of the first Mk Is, the SAAF's dormant aerobatic team was resuscitated, giving its first display at the opening ceremony of Atlas Aircraft Corporation in late 1967. The unit was dubbed the Silver Falcons in 1970, and its aircraft were painted in orange, white and blue to represent the South African flag of the time. (SAVA)

59. Sud Aviation/Aérospatiale Puma

Designed by France's Sud Aviation, the twin-engined Puma was developed to meet the French Air Force's need for a medium-sized, all-weather tactical helicopter. A new SAAF squadron, No. 19, was formed at AFB Swartkop to accommodate the new helicopters. (Col André Kritzinger, CC BY-SA 3.0, Wikimedia Commons)

Late in 1969, South Africa became one of the first buyers of the French-made Puma when it acquired 20 (Nos. 121–140) of the SA.330C version. This was followed in January 1975 by the purchase of 18 (Nos. 141–160) of the newer SA.330H variant, which were all assembled at Snake Valley near AFB Swartkop, Pretoria. Then in November 1977, the SAAF started to receive its final Pumas, this time 29 (Nos 161–190) of the SA.330L variant. (Col André Kritzinger, CC BY-SA 3.0, Wikimedia Commons)

From the beginning of 1972, the Puma commenced counter-insurgency operations in South West Africa and Angola, being employed on rapid troop deployment, casualty evacuation, radio relays and aircrew rescue. Four Pumas were brought down by enemy ground fire during the Border War. (GHR_ZA, CC BY-SA 3.0, Wikimedia Commons)

60. Piaggio Albatross

The high gull wings and twin backward-facing Lycoming engines make for easy identification of the Italian Piaggio P.166, officially named the Albatross by the SAAF. The first Albatrosses arrived by ship in Cape Town in early 1969, where they were assembled and delivered to SAAF No. 27 Squadron at AFB Ysterplaat. (Farawayman, CC BY-SA 3.0, Wikimedia Commons)

Entering service on short-range coastal patrol and search-and-rescue duties, shortly after receiving 20 radar-equipped versions, in the mid-1970s the squadron started to operate out of what was then the D.F. Malan International Airport, Cape Town. The Albatrosses were retired in 1993, when two aircraft were allocated to the SAAF Museum Historic Flight. (SAVA)

61. Transall C-160

A 1959 joint venture between French and German aviation companies — Transporter Allianz — resulted in the design of a twin-engine general transport, the Transall C-160. Powered by two Rolls-Royce Tyne turboprops, the C-160 has a payload of eight tons. In addition to a crew of four, it can accommodate 90 troops or 64 fully kitted-out paratroops. (Aeroprints, CC BY-SA 3.0, Wikimedia Commons)

After pilot conversion training in France, SAAF No. 28 Squadron, based at AFB Waterkloof, took delivery of the first of nine C-160Zs in August 1969. Designed for tactical transport work, the C-160 had a robust capability to take off and land in incredibly short distances. (SAVA)

The C-160 proved invaluable in the Border War, particularly in the deployment of paratroops, the conveyance of casualties back to South Africa and the high-risk delivery of war matériel to UNITA forces during the Angolan civil war (1975–2002). However, the C-160 was a costly aircraft to maintain, the French charging exorbitantly for spares and maintenance. Eventually, eight mothballed aircraft were sold as scrap iron. (paulopanz)

62. Aermacchi Bosbok

Its roots in 1960s Mexico, the Italian firm of Aermacchi became the exclusive licensed manufacturer (except the USA) of the AM.3. Whilst using the wing design of the Aermacchi AL-60 utility aircraft, strengthened to incorporate two hardpoints, the fuselage was a new design. The third prototype used a more powerful Piaggio-built Lycoming engine. This variant was designated AM.3C. (Andries van Tonder, www.facebook.com/affordablememoriesphotographers)

From March 1973 to December 1975, SAAF Nos. 41 and 42 Squadrons took delivery of 40 of this variant, referred to in South Africa as the Bosbok. Seeing extensive action in the Border War, the Bosbok could carry 375lb of ordnance under each wing, including rockets, bombs and missiles. (Phil Vabre, GFDL V 1.2, Wikimedia Commons)

63. Hawker Siddeley Mercurius

In March 1970, the SAAF's No. 21 (VIP) Squadron, based at AFB Swartkop, took delivery of the first of four British Hawker Siddeley HS.125-400B mid-size business jets, named Mercurius by the SAAF. On 26 May 1971, in a tragic accident, a formation of three Mercuriuses crashed into Cape Town's Devil's Peak, killing all 11 air force and army personnel on board. (SAAF)

Originally designed by de Havilland and initially designated as the DH.125 Jet Dragon, the type entered production in 1963 as the Hawker Siddeley HS.125. The RAF was a significant buyer, operating the aircraft in airborne training and air force navigation roles. When Hawker Siddeley acquired de Havilland before the start of the project, the Series 4, which featured numerous minor refinements, was marketed as the Series 400A and 400B and 116 were built. The 400B variants acquired by the SAAF, named the Mercurius, were powered by two Rolls-Royce Viper 522 turbojet engines. The SAAF's Mercurius fleet was officially withdrawn from service in 1998 and later sold. In 1977, Hawker Siddeley Aircraft merged with the British Aircraft Corporation to form British Aerospace, and the type designation was changed to BAe 125. (SAAF)

64. Atlas Kudu

Late in 1971, South Africa made the decision to locally design and manufacture the Kudu. Utilising the same engine, wings and numerous associated parts as the Bosbok (see page 79), the resultant C-4M 'Kudu' was specifically tailored to operate in African conditions. Developed at the Atlas factory near Johannesburg, the Kudu featured a single sliding door to deploy paratroopers, and a hatch in the floor for airdrops. (SAVA)

The Kudu went into service with SAAF Nos. 41 and 42 Squadrons in 1974, then flying out of AFB Swartkop. The Kudu was an extremely useful utility asset, moving personnel and light cargo around the Border War operational areas. Able to land at most bases, the Kudu needed only 260 metres, often with way over its payload limit of 650kg or seven passengers. Kudus also flew spotter, route-reconnaissance and communications missions. (Alan Wilson, CC BY-SA 4.0, Wikimedia Commons)

65. Dassault Mirage F1

Successor to the Mirage III, the F1 was significantly different to its antecedent. Designed as an all-weather, fighter/interceptor, the F1 featured swept wings and the powerful Snecma TF306 turbojet engine. Arriving first in April 1975, 16 of the F1CZ variant were allotted to SAAF No. 3 Squadron, AFB Waterkloof. Later that year, 32 F1AZs were received and given to SAAF No. 1 Squadron, also based at Waterkloof before moving to AFB Pietersburg in 1981. (Andries van Tonder, www.facebook.com/affordablememories-photographers)

The F1 is armed with two 30mm DEFA 553 cannon in the forward fuselage, while underwing and ventral hardpoints accommodated external armament. In the interceptor role, each wingtip could be fitted with Matra R.550 Magic 1 air-to-air missiles. Adding to this configuration are two underwing Matra R.530 air-to-air missiles. In the ground-attack role, underwing multiple-tube rocket launchers may be fitted. (SAVA)

In November 1978, five F1CZs were deployed to AFB Ondangwa in South West Africa (now Namibia) to provide escort to SAAF reconnaissance flights over southern Angola. In a back-up role to the F1AZ, they were armed with Matra M155 rocket pods or 250kg bombs to conduct pre-emptive strikes against insurgent targets. (Col André Kritzinger, CC BY-SA 3.0, Wikimedia Commons)

66. Swearingen Merlin

The American Swearingen (Fairchild) Merlin IVa is a seven-passenger, all-weather, pressurised executive-type transport. First acquired in 1975, the aircraft are operated by SAAF No. 21 (VIP) Squadron out of AFB Swartkop. One of the Merlins is a bespoke air ambulance with the most modern medical equipment. (Ray W)

The Merlin was developed by Ed Swearingen of Swearingen Aircraft from 1965 onwards, and produced by Fairchild, which had taken over the company in 1971. The SAAF acquired seven of the IVA variant, powered by two Garrett AiResearch TPE-331 turboprop engines, and the first arrived at Rand Airport in June 1975. All four carried an Aerojet General rocket installation in the tail cone, designed to assist in an emergency, if required. The SAAF Merlins also featured a large cargo door in the rear fuselage, while passengers used the front door. On 14 July, SAAF Merlin SA226 (ZS-JLZ) and a Piper PA-31 Navajo (ZS-KTX) crashed south of Pretoria, South Africa, following a mid-air collision. The Merlin was returning from Namibia and was descending on to a wide, left-hand base-leg to land at AFB Waterkloof when the aircraft collided. The Navajo had just taken off from Lanseria Airport and was in the climb on a flight to Wonderboom Airport. Five members of the SAAF, including two major-generals, a colonel from the Danie Theron Combat School and six civilians, including two young boys, perished in the tragedy. (SAAF)

Chapter 3
Squadrons

No. 1 Squadron
Formed at Swartkop AFS in February 1920, the squadron was equipped with de Havilland DH.9 biplanes. After moving to AFB Waterkloof, in February 1940, it was re-equipped with four Hawker Hurricanes and six Hawker Furies, at the time the most modern fighters in the SAAF.

The squadron entered World War Two service against the Italians in East Africa in May 1940, before joining the Allies in North Africa where, for the rest of the war, it operated Mk V, Mk VIII and Mk IX Supermarine Spitfires.

In 1950, the squadron was equipped with de Havilland Vampires, which were replaced with Canadair Sabres in September 1956. Late in 1963, the squadron became the SAAF's sole Sabre operator. Upon its move to AFB Pietersburg in 1967, Atlas Impalas were added to the fleet. The squadron returned to Waterkloof in 1975, where it was re-equipped with the Dassault Mirage F1AZ.

No. 2 Squadron
The squadron started off in January 1939 as No. 2 (Transvaal) Squadron at AFB Waterkloof where it was equipped with Hawker Hartebees. Converting to Hawker Furies and Hurricanes and Gloster Gladiators, the squadron was re-formed in East Africa in October 1940.

In April 1941, the squadron moved to Egypt where, now equipped with Curtiss Tomahawks, it entered the Western Desert theatre three months later against the Luftwaffe. Here the illustrious career of the famed "Flying Cheetahs" had a solid footing. By May 1942, the squadron started receiving Curtiss Kittyhawks. With Allied victory over the Germans and Italians in the North African Campaign, the squadron, now re-equipped with Spitfires in a ground-attack role, participated in the invasion of Sicily and Italy.

Disbanded at the end of the war, the squadron was re-formed at AFB Waterkloof at the end of 1948. From 1950 to 1953, the squadron saw distinguished service in the Korean War, first flying North American F-51 Mustangs and then North American F-86 Sabres. After the conflict, the squadron converted to de Havilland Vampires, and in 1956, Sabre 6s. Then, in 1963 the squadron received Dassault Mirage IIIs, assuming the fighter-reconnaissance and fighter-bomber roles.

No. 3 Squadron
Equipped with Hawker Hartebees and Hurricanes, the squadron was formed at AFB Waterkloof in September 1940. A month later, it was deployed to the East African theatre. In December 1942, the squadron was reformed in South Africa, equipped with Hurricanes and Spitfires, and sent to the Middle East. From here, now with Spitfire IXs, the squadron joined the new SAAF No. 8 Wing in Italy.

As was the case with many SAAF squadrons, the unit was disbanded at the end of the war before being re-formed at Baragwanath Airport as a citizen, part-time unit, flying North American Harvards. Seven years later, the squadron was axed once more, only to be re-formed in August 1966 at Waterkloof. Essentially only a paper unit under No. 2 Squadron, the unit had Dassault Mirage IIIs at its core. It reverted to being an autonomous squadron in February 1975 when it received its first Dassault Mirage F1CZs.

No. 4 Squadron
The squadron was first formed at Durban in April 1939 and equipped with Curtiss Mohawks and Hawker Furies, but after only eight months it was disbanded. In March 1941, the squadron was resuscitated and underwent operational training in East Africa before being moved to Egypt for conversion to Curtiss Tomahawks. After seeing action in North Africa from November 1941, the squadron, now equipped with Curtiss Kittyhawks, moved on to Italy where, ultimately equipped with Spitfires, it saw out the war.

Disbanded in October 1945, the squadron was re-formed as an Active Citizen Force, its strength drawn from No. 1 Squadron. After being disbanded once more in October 1958, the squadron was brought to life for a third time at Swartkop in November 1961, and equipped with North American Harvards. In 1972, the squadron converted to Impala Mk Is, and in November 1976 it received its first Impala Mk IIs.

No. 5 Squadron
Named after the great Zulu warrior-king, the 'Chakas', the squadron's first incarnation lasted only eight months after its inception as a fighter-bomber unit in Cape Town in April 1939. After being re-formed at Swartkop AFS in May 1941 and trained on Curtiss Mohawks, the squadron was deployed to Egypt that December where it flew Curtiss Tomahawks on shipping patrols. Responding to a greater need, the squadron was transferred to North Africa where it adopted a fighter role. Converting to Curtiss Kittyhawks towards the end of 1942, the squadron took on extra ground-attack duties. With the defeat of Axis forces in North Africa, the squadron was deployed to Sicily, mainly for ground attacks on German anti-aircraft positions. In October 1943, the squadron moved to Italy where it assumed close-support and fighter-bomber duties.

Disbanded at the end of the war, the squadron was re-formed in Durban in December 1950, also as an Active Citizen Force unit flying Harvards. Receiving the freedom of the city of Durban in 1970, three years later the squadron converted to Atlas Impalas.

No. 6 Squadron
Although designated a 'shadow' fighter unit when it was formed in Cape Town in April 1939, the squadron was equipped with Westland Wapitis engaged in anti-submarine coastal patrols. In February 1940, the unit was renumbered No. 1 Squadron. Two years later, a new No. 6 Squadron was created at Swartkop AFB and equipped with Curtiss Mohawks. Deployed to Natal when a Japanese invasion threatened but never materialised, the squadron was moved to Eerste River in the Western Cape, where it took on a number of obsolete Westland Wapitis, Hawker Furies and Fairey Battles. Disbanded in July 1943, the squadron was reactivated as an Active Citizen Force in July 1952. Based at Port Elizabeth, the squadron flew North American Harvards. Then, following a disbandment from 1959 to 1961, and still flying Harvards, the squadron was reformed. In March 1975, Atlas Impalas were added to the fleet.

No. 7 Squadron
Formed as a fighter unit at AFB Swartkop in January 1942, the squadron was equipped with North American Harvards and Curtiss Mohawks. Posted to Egypt in May 1942, the squadron converted to aged Hawker Hurricane Mk Is, restricting its operations to meteorological and local-defence flights. The unit was then attached to No. 7 Wing SAAF in the Western Desert where it flew top cover for RAF Hurricanes. In September, after a spell of well-earned rest at AFS Shandur, the squadron converted to Hawker Hurricane IIds, which required specialist training in the operation of the aircraft's two 40mm 'tank-buster' cannon. Paired at El Alamein with No. 6 Squadron RAF in the Desert Air Force's 'Flying Can Openers' of No. 211 Group, the squadron executed ground-attack missions behind enemy lines. In January 1943, the squadron was based at Benina for a brief spell, flying Hawker Hurricanes and Supermarine Spitfires on coastal reconnaissance patrols. After the end of hostilities in North Africa, the squadron was attached to No. 212 Group, Air Defence Eastern Mediterranean. In April 1944, the unit joined No. 7 Wing in Italy where duties included fighter-bomber missions, armed reconnaissance and bomber escort.

When the war in Europe ended, the squadron was despatched to the Far East where Japan was yet to surrender, but it was turned around at Ceylon (now Sri Lanka) when the war in the Pacific stopped. Returning home for disbandment in September 1945, the squadron was re-formed at AFB Ysterplaat in 1951. After eight years, and still flying North American Harvards, the squadron went into mothballs for two years. After being re-activated at Youngsfield, near Cape Town, in August 1961, the squadron moved to a new home at Ysterplaat in October 1969. Now classed as a Citizen Force, in 1977 the squadron received Atlas Impalas and moved to the D.F. Malan International Airport, Cape Town (now O.R. Tambo).

No. 8 Squadron
One of the shortest-lived World War Two units of the SAAF, the squadron was formed in February 1942, flying Hawker Furies. However, before becoming operational, the squadron was disbanded in August that year. It would only be in 1951 that the squadron was re-formed, this time as an Active Citizen Force, flying Harvards at AFS Bloemspruit. In the mid-1970s, the squadron received Atlas Impala Mk I and Mk II jet aircraft.

No. 10 Squadron
From its formation at East London in April 1939 until its disbandment in 1943, the fighter-bomber unit was deployed on South African coastal patrols. Initially equipped with Hawker Furies, the squadron converted to Curtiss Mohawks and Kittyhawks at the time of a Japanese invasion scare. Disbanded in July 1943, the unit was re-formed in May 1944 and moved to Syria for training as a fighter squadron flying Supermarine Spitfires. After being based in Libya and flying operational sorties over the Mediterranean, the squadron was mothballed in October 1944.

As a postscript, the squadron was brought back to life in 1986 at AFB Potchefstroom and equipped with the Kentron Seeker unmanned aerial vehicle in the role of reconnaissance and artillery weapons-delivery guidance support.

No. 11 Squadron
Based at Durban as a fighter-bomber squadron, the unit was raised in April 1939, and in that December, after a brief disbandment, was re-formed at AFB Waterkloof by the renumbering of No. 1 Squadron. In May 1940, the squadron was equipped with Hawker Hartebees and deployed to Kenya for armed reconnaissance on the colony's Northern Frontier. Later that year, the unit converted to Fairey Battles, as it continued in the same role throughout the campaign in Italian East Africa. As the campaign wound down, in May 1941 the squadron was renumbered No. 15 Squadron. Equipped with Supermarine Spitfires, the squadron was resuscitated in the Middle East in June 1944. Deployed to Italy after training, the squadron converted to Curtiss Kittyhawks. Mothballed at the end of the war, the unit was re-formed in January 1974 at AFS Potchefstroom, where it was equipped with Cessna 185s.

No. 12 Squadron
Shortly after the outbreak of World War Two, No. 2 Squadron at AFB Waterkloof was retitled No. 12 Squadron. After starting off with aging Avro Ansons, in 1940 the squadron acquired ten civilian Junkers Ju 86s from the national airline, South African Airways. Following a rapid conversion, in May the squadron was deployed to Kenya for bombing duties. In March 1941, a detachment of Martin Marylands from Nos. 14 and No. 60 SAAF Squadrons was attached on a temporary basis in a reconnaissance role. With an attachment of B-25 Mitchells of the US Air Force, the squadron, now equipped with Douglas Bostons, participated in the battle of El Alamein in 1942. In a build-up to the Allied invasion of Sicily, in May 1943 the squadron commenced air reconnaissance and bombing sorties over the island. In October, the squadron moved onto mainland Italy. By December, the squadron had converted to Martin Marauder Mk IIs. Throughout 1944 and until the end of the war in Europe, the squadron rendered valuable air support to the British Eighth Army during the offensive to push German forces out of Italy.

The squadron's war service swansong before disbandment was to assist with the repatriation by air of the South African Eighth Armoured Division. A year later, re-equipped with Avro Ansons and, later, Sikorsky S-51 helicopters, the squadron undertook two years of anti-tsetse fly aerial spraying. It was then disbanded before being reactivated in 1963 as the SAAF's first jet bomber squadron, flying English Electric Canberras.

No. 15 Squadron

This was another squadron thrown together in Germiston at the outbreak of World War Two, the outbreak of which found South Africa wholly unprepared. After being transferred to Wingfield on long-range maritime duties with only three Junkers Ju 86s, in December 1939, the squadron was absorbed as A Flight of No. 32 Squadron. In June 1941, the squadron was re-activated and spent two months in East Africa flying tired Fairey Battles inherited from No. 11 Squadron. The unit was then ordered back to Pretoria for conversion training on Martin Maryland light bombers. Early in 1942, the squadron moved to Egypt, where it was converted to Bristol Blenheim bombers and employed on reconnaissance and ground-attack operations. From January to July 1943, the squadron was based on Cyprus, from where it was engaged in anti-submarine patrols in the Eastern Mediterranean. Now equipped with Martin Baltimores, the squadron conducted patrols, shipping reconnaissance and bombing raids in the Aegean until August 1944 when it joined Allied forces in Italy.

At the end of the war in Europe, the squadron was disbanded, and would only re-appear in 1968 when it was re-formed at AFB Swartkop. It would be the first SAAF unit to be equipped with the Aerospatiale Super Frelon, the SAAF's largest helicopter. Since then, the squadron has flown food- and flood-relief sorties, and became a stalwart of the Border War.

No. 16 Squadron

Another World War Two product, the squadron's humble beginnings were at Germiston in September 1939, from where it was posted to Walvis Bay to patrol the South West African coast with a paltry three Junkers Ju 86s acquired from the national airline. In mid-1941, the squadron found itself at Addis Ababa in Italian East Africa, where a further five Ju 86s were added to the fleet, together with two Martin Marylands and two Fairey Battle light bombers on loan. In an unexpected move, and after considerable action in East Africa, in August 1941 the squadron was disbanded. In September of the following year, the squadron was re-formed with the renumbering of No. 20 Squadron. Assigned to assist in the East Africa Command's invasion of Madagascar, the squadron was equipped with Martin Marylands and Bristol Beauforts in support operations against Vichy French forces on the island. At the end of the campaign, the squadron converted to Bristol Blenheims, also known as Bisleys. In April 1943, the squadron was moved to Egypt on anti-submarine patrols. Here the Blenheims were replaced with Bristol Beauforts. In November, the fleet was upgraded to the formidable Bristol Beaufighter with its powerful armament of cannon and rockets. Early in 1944, the squadron commenced anti-shipping sorties in the Aegean before joining the Balkan Air Force in support of the Yugoslav partisans.

Disbanded in June 1945, the squadron had earned an impressive eight battle honours. Reformed at AFB Ysterplaat early in 1968, the squadron now flew Aerospatiale Alouette III helicopters. In January 1969, the unit moved to Durban until mid-1972 when it moved to Bloemfontein. At this time, the squadron was divided into two flights: A at Port Elizabeth and B, remaining at Bloemfontein. In December 1980 the two flights were rejoined at Port Elizabeth.

No. 17 Squadron

The squadron's initial existence was short-lived. Formed at Swartkop AFS in September 1939, and equipped with former South African Airways Junkers Ju 52/3ms, barely three months later it was combined with Nos. 18 and 19 Squadrons as No. 50 Squadron. It would only be in October 1942 when the squadron was re-activated at Mobile Air Force (MAF) Depot, Voortrekkerhoogte, as a general reconnaissance unit. In January 1943, it moved to Aden in the Middle East for conversion to Bristol Blenheims and convoy escort duties. The squadron would remain in the Middle East and North Africa, during which time it converted to Lockheed Ventura light bombers. After undergoing conversion training on Vickers Wellingtons and Warwicks, the squadron returned to South Africa in September 1945, and was disbanded six months later. However, this proved to be a brief spell in mothballs, and the squadron was brought out of retirement in June 1947 when it became known as the City of Cape Town Squadron.

After a two-year period back on the shelf, in December 1957, the squadron was re-formed at Langebaan as the SAAF's only exclusively helicopter squadron, operating Sikorsky S-55s and an S-51. In December 1960, the squadron took delivery of its first Aerospatiale Alouette IIs. From here, the unit moved to AFB Ysterplaat, where it received Aerospatiale III helicopters and split into four flights. A and B Flights were based at Pretoria and Bloemfontein respectively, while C and Training Flights remained at Ysterplaat. C Flight would eventually form the nucleus of the reconstituted No. 16 Squadron and the Training Flight became the Helicopter Conversion Unit.

No. 19 Squadron

Like some of the other wartime squadrons, this unit was hastily formed at AFB Swartkop at the outbreak of hostilities, equipped with Junkers Ju 52/3ms, and disbanded three months later. The squadron then reappeared in September 1944 when No. 227 Squadron RAF was taken over by the SAAF in Biferno, Italy. Operating Bristol Beaufighters, the squadron went into immediate service attacking German lines of communication in Yugoslavia and Greece. At this time, the 60lb RP-3 rocket was supplied to the squadron, which significantly bolstered its arsenal in support of Yugoslavian partisans. Disbanded after a distinguished wartime record, the squadron had its number reused in the early 1970s to form a new unit from a flight of No. 17 Squadron Aerospatiale Pumas. The squadron became a key player in the Border War.

No. 21 Squadron

Prior to becoming the SAAF's principal VIP transport, this squadron led a very active war existence characterised by bold bombing raids and air-to-air battles. After formation in Kenya in May 1941, the squadron entered the Western Desert theatre flying Martin Marylands. Now the last unit in the region still operating Marylands, the squadron was withdrawn to the Nile Delta in January 1942. A period of non-operational duties ensued as the squadron prepared to take delivery of Martin Baltimores. Attached to No. 232 Wing SAAF in October 1942 and then No. 3 Wing SAAF, the squadron fought in the Battle of El Alamein and eventually participated in the final bombing sortie of the North African campaign. However, there was no letup for the unit, as it fought its way on to Sicily via the islands of Pantelleria and Malta. In October 1943, the squadron found itself at Foggia AFB on mainland Italy.

Disbanded in Italy in September 1945, the squadron was re-activated a year later at AFB Swartkop as a bomber unit flying Lockheed Venturas. However, early in January 1951, the unit was renumbered No. 25 Squadron. Having identified the need for a VIP unit, the SAAF re-formed the squadron in February 1968. Forming part of No. 28 Squadron, the unit flew a Vickers Viscount and three Douglas Dakotas. From 1970, the squadron started to receive the Hawker Siddeley HS.125 Mercurius business jets.

No. 22 Squadron
With a history of close co-operation with the South African navy, the squadron was formed in Durban in July 1942 as a coastal reconnaissance unit, air-sea rescue, convoy escort and anti-submarine unit. Equipped initially with converted Junkers Ju 86 airliners and near-obsolete Avro Ansons, the squadron took delivery in August of eight Lockheed Venturas to replace the Junkers. Until July 1944, when the squadron moved to Gibraltar, its operations were confined to the Indian Ocean area. Upon the cessation of hostilities, the squadron was based in Egypt until its disbandment in October 1945.

The squadron was re-formed in 1954, equipped with Venturas, but later disbanded. With the acquisition of Westland Wasp helicopters, the SAAF resuscitated the unit as No. 22 Flight in January 1964. With additional Wasp helicopters, the unit received full squadron status in May 1976. The squadron also operates Aerospatiale Alouette III helicopters.

No. 24 Squadron
Formed in March 1941 by retitling No. 14 Squadron, which had been on active service in East Africa, the squadron was equipped with Martin Maryland light bombers. These, in turn, were replaced with Douglas Bostons when the unit was sent to North Africa. From December 1943 onwards, the squadron operated Mk II and Mk III Martin Marauders. It was attached to SAAF No. 3 Wing at the end of the war, and the Marauders were converted to transports.

Five months after its disbandment, the squadron was resuscitated at AFB Bloemspruit in April 1946 as a bomber unit equipped with Lockheed B-34 Venturas. Disbanded for a second time in January 1951, when the squadron's Active Civilian Force created the core of No. 8 Squadron, it would be another 14 years before the squadron was re-formed, this time at the Royal Navy Air Station, Lossiemouth, Scotland. The SAAF had acquired Hawker Siddeley Buccaneers, so pilots from the squadron were sent to Lossiemouth for conversion training and to fly the new assets back to AFB Waterkloof.

No. 25 Squadron
Officially classed as a bomber, reconnaissance and torpedo squadron, the unit was formed at Cemetery Camp, Port Elizabeth, in July 1942, with a handful of near-obsolete Avro Ansons and a vintage Westland Wapiti. In February 1943, now flying Lockheed Venturas, the squadron was rebased at St Albans, also in the Port Elizabeth area. Consolidated Catalinas of the Dutch navy were now attached to operate night patrols while the squadron's Venturas performed that task during the day. By mid-1944, the squadron had been deployed to Italy, where it was re-issued with B-34 versions of the Ventura. Operating against shipping and ports along the Dalmatian coast, in November the squadron added Martin Marauders to its assets, which allowed for support operations over Yugoslavia.

Disbanded at war's end, in January 1951 the squadron was reformed as a part-time unit flying Douglas Dakotas. It was retitled No. 44 Squadron in 1953, before being re-formed in February 1968 at AFB Ysterplaat, still flying Dakotas.

No. 27 Squadron
Officially designated a TBR (torpedo, bomber and reconnaissance) squadron, the maritime flight was formed at Eerste River, near Cape Town, in August 1942. Operating Lockheed Venturas, the unit went operational along South Africa's west coast early in 1943. In June 1944, the squadron commenced operations in the Mediterranean and along the Spanish coast, but as the theatre became quieter, in November the squadron started to move back to South Africa to its new home at AFS Swartkop. A small detachment remained in Egypt for conversion training to Vickers Wellingtons and Warwicks.

At the conclusion of the war, the squadron remained in the Middle East on air, sea and rescue duties, before returning to South Africa, where it was disbanded in December 1945. Reformed at AFB Ysterplaat in 1951 as a part-time maritime unit flying Venturas, after seven years the squadron was disbanded before being reconstituted in October 1962, flying Dakotas. In 1969, the squadron converted to Piaggio Albatrosses on medium-range maritime patrol duties.

No. 28 Squadron
The squadron has been a transport unit ever since its formation in Egypt in June 1943. The fleet comprised two Vickers Wellingtons, five Douglas Dakotas and seven Avro Ansons. At the end of the war, the squadron remained in service, based at Swartkop AFS, from where it operated an essential shuttle service bringing South Africans back home from overseas. VIP flights remained an integral function of the squadron, and over time the Ansons were replaced with bigger and more modern aircraft. In 1949, nine de Havilland Devons, a version of the DH.104 Dove, were acquired for the VIP flight, and in 1955, two de Havilland Herons were added. The Venturas were phased out in the late 1950s, and a Vickers Viscount added to the VIP fleet. In the early 1960s, the Herons and Devons were mothballed. This was followed by the acquisition of the Lockheed C-130 Hercules transports, and the shedding of the VIP element to No. 44 Squadron. In 1969, the squadron took delivery of the Transall C-160.

No. 30 Squadron
In August 1944, at Pescara in Italy, the squadron was born from the renumbering of No. 223 Squadron RAF. Taking over Martin Baltimores, virtually straightaway the squadron converted to Martin Marauders. Operating from Pescara and Jesi, in June 1945 the squadron moved to Biferno where it was disbanded the following month. In 1980, the squadron was reactivated as the SAAF's second Aerospatiale Super Frelon unit.

No. 31 Squadron
In December 1939, Nos. 13 and 14 coastal bomber/reconnaissance squadrons SAAF were amalgamated to form this squadron. The unit's aging Junkers Ju 86s and Bristol Blenheims were soon replaced with Avro Ansons, and in September 1940, the squadron was split into No. 31 Coastal Reconnaissance Flight, flying from Durban, and No. 33 Coastal Reconnaissance Flight, flying out of Port Elizabeth. Re-activated at AFB Swartkop In January 1944, the squadron was deployed to Lydda In Palestine for conversion training on Consolidated B-24 Liberator bombers. Joining No. 34 Squadron under No. 2 Wing SAAF at Kilo 40 camp north of Cairo in May 1944, the unit conducted bombing raids over Crete and the Aegean. In June, the squadron was attached to No. 205 Group RAF operating from Celone, near Foggia in southern Italy. Over time, operations were expanded to include strategic targets in the Balkans, Hungary, Austria, Romania and northern Italy. The squadron also conducted air supply drops to Yugoslav partisans as well as mine-laying missions on the Danube River. The squadron then participated in Operation *Dragoon*, the invasion of southern France in August 1944. This was followed by long-range missions to Poland, transporting war matériel to the resistance. During the four-week operation, 31 of the 80 aircraft used were shot down, of which 25 were SAAF Liberators. In total, 69 South Africans lost their lives. At the end of the war in Europe, the squadron converted to transport duties until it was disbanded in December 1945. The squadron was re-formed in 1982 as a helicopter unit.

No. 35 Squadron
Formed in February 1945 by renumbering No. 262 Squadron RAF, during World War Two the squadron had operated Consolidated Catalinas on the South African and South West African (now Namibian) coasts. In April 1945, the squadron took delivery of 16 Short Sunderlands which, too late for active service, were employed to ferry South African troops from Egypt home to South Africa. From 1948 to 1950, an Active Citizen Force element was added which flew Harvards, Venturas, Oxfords and Spitfires. In 1957, the Sunderlands were replaced by Avro Shackletons, and the squadron moved to Cape Town.

No. 40 Squadron
Emerging from World War Two with one of the most distinguished combat records in the whole of the SAAF, the squadron was first formed at AFB Waterkloof in May 1940. Equipped with Hawker Hartebees, this army cooperation tactical reconnaissance unit moved to East Africa almost immediately. In August 1941, the squadron was brought home to prepare for deployment to Egypt in November. Familiarisation of pilots' role in desert conditions was achieved by attachment to existing operational Royal Air Force and Royal Australian Air Force tactical reconnaissance units. In February 1942, the squadron received Hawker Hurricanes and Curtiss Tomahawks and went operational in March. After participating in the Battle of El Alamein in October, from February 1943 the squadron received camera-equipped Supermarine Spitfires to assist with the final Allied victory in North Africa. However, there would be no rest for the squadron's air and ground crews, and the unit set a record for the number of sorties flown during the Allied invasion of Sicily in June. In September, leading elements of the squadron moved onto the Italian mainland, while the rest of the squadron flew armed reconnaissance and tactical sorties for the invading British and American armies on the ground. Then, even as the conflict drew to an end, in March and April 1945 the squadron recorded its busiest time of the war. When Nazi Germany surrendered, the squadron was the only remaining unit of the Allied Desert Air Force still operational. Finally, in October, the gallant but exhausted squadron returned home to be disbanded. Re-formed at AFS Bloemspruit in January 1951, the squadron was equipped with North American Harvards before moving to Dunnotar.

No. 41 Squadron
Initially formed as an 'army cooperation' squadron at AFB Waterkloof in October 1940, the unit was deployed to Italian East Africa to fly Hawker Hartebees on photo-survey missions, artillery spotting and dropping propaganda leaflets. However, by December it had also adopted an offensive role, strafing and bombing ground targets. By early 1942, the squadron participated in communications flying and mopping-up operations of remaining Italian resistance. From mid-1943, it moved north for convoy patrol and air defence duties in North Africa and parts of the Mediterranean. The squadron converted to Hawker Hurricanes, and in February 1944, took delivery of its first Supermarine Spitfires to use for long-range escorts of No. 24 Squadron Martin Marauder missions over Crete. The unit was disbanded in October that year and its personnel dispersed to other SAAF squadrons.

Re-formed in 1963 as the first part-time army co-operation unit, in 1968 it returned to the SAAF, flying Cessna 185s from Grand Central Airport, Johannesburg. In 1974, the squadron was equipped with the Aermacchi/Atlas Bosbok, and in 1976, the Atlas Kudu.

No. 42 Squadron
Uniquely, the squadron's immediate antecedent, No. 42 Air Observation Post (AOP) Flight, was the only South African army aviation unit of World War Two. Formed at Bari in Italy early in 1945, the flight, equipped with Austers, played an important role in directing the artillery of the 6th South African Armoured Division as the British Eighth and American Fifth Armies finally forced the capitulation of German forces in Italy.

After the war, the flight and its Austers remained with the army, and was based at the South African Artillery in Potchefstroom. After a spell at the SAAF's Central Flying School, the flight reverted to the army in 1953. After a few years, squadron status was granted, and No. 42 Squadron found a permanent home in the SAAF. In 1962, the Austers were replaced with Cessna 185s, and in 1974 by Atlas Bosboks.

	**No. 44 Squadron** Created in Egypt in March 1944 by the renumbering of No. 43 Squadron, this transport unit first flew Avro Ansons. Within a month, with assistance from the Royal Air Force and No. 28 Squadron SAAF, its members saw intensive conversion training to Douglas Dakotas. Scheduled cargo and passenger runs commenced with Middle East destinations, before spreading as far afield as the Ivory Coast (now Ghana) and India. In February 1945, the squadron moved to Bari, Italy, primarily for operations in the Balkans. High-risk night drops on rustic airstrips behind enemy lines supplemented daytime airdrops to Yugoslav partisans. Disbanded in December 1945 while still on active service, in November 1952 the squadron was re-activated as an Active Civilian Force Dakota transport flight at AFS Swartkop by retitling No. 25 Squadron. In 1966, Douglas DC-4s were added to the fleet.
	**No. 60 Squadron** From its creation in Nairobi, Kenya, in December 1940, the squadron flew the British Aircraft Double Eagle, Martin Maryland, de Havilland Mosquito and the Lockheed Ventura. In 1943, the squadron was attached to the North African Photo Reconnaissance Wing, which later became the Mediterranean Allied Photo Reconnaissance Wing.

The Silver Falcons is the official aerobatic team of the SAAF. Since 1967, 350 displays have been performed with the Impala Mk I (pictured), before it was replaced with the Pilatus PC-7 Mk II. Today, the team is based at the Central Flying School, AFB Langebaanweg. (surclaro)

Chapter 4

Into the Future

The Border War in Namibia and Southern Angola continued until 1989, when the withdrawal of the South African Defence Force paved the way for the former's independence in 1990. The largely unconventional war saw virtually every SAAF fixed-wing aircraft and helicopters in operation at some stage. However, sophisticated ground and air threats to the SAAF placed increased pressure on the government to upgrade fleet capabilities. Angolan 9K31 Strela surface-to-air missile systems were captured by South African ground forces, while SAAF pilots increasingly encountered MiG-21s of the Cuban Air Force.

This prompted South Africa to task the Atlas Aircraft Corporation (later Denel Aeronautics) to design and produce the 'Cheetah' as a major upgrade of the French-built Dassault Mirage III fleet operated by the SAAF. The programme employed state-of-the-art technology from the Israeli-built Kfir, an all-weather, multirole combat aircraft based on the French Dassault Mirage 5.

Other developments included the unveiling of the new Darter V32C air-to-air missile and the Denel Dynamics Seeker unmanned aerial vehicle. Towards the late 1980s, work commenced on the 'Oryx' helicopter, an upgraded and remanufactured version of the Puma, essentially equivalent to the Eurocopter AS332 Super Puma.

According to André Wessels, in his paper 'The South African Air Force, 1920–2012: A Review of its History and an Indication of its Cultural Heritage' (*Scientia Militaria* vol 40, no 3, 2012, pp.222-249.

Now retired, Cheetah E No. 826 was a local conversion of the Mirage IIIEZ. It is on display in the Sci-Bono Discovery Centre, Johannesburg, South Africa. (Alan Wilson, CC BY-SA 4.0, Wikimedia Commons)

The highly manoeuvrable Swiss Pilatus PC-7 Mk II, in its distinctive blue and white livery and tail in the national colours, replaced the Impalas operated by the SAAF's official aerobatic team, the Silver Falcons. (Col André Kritzinger, CC BY-SA 3.0, Wikimedia Commons)

In 1998, the SAAF took delivery of the problem-beset, highly sophisticated Rooivalk 'tank-buster' attack helicopter, manufactured in South Africa by Denel Aeronautics. The company signed an agreement with Turkish defence corporation Aselsan in 2023, to collaborate on the avionics upgrade of the Rooivalk. (Col André Kritzinger, CC BY-SA 3.0, Wikimedia Commons)

doi: 10.5787/40-3-1043), the strength of the SAAF in 1989 stood at 801 fixed-wing aircraft and 24 helicopter types. This included 43 Mirage F1s, 50 Cheetahs/Mirage IIIs, five Buccaneers, seven Canberras, 212 Impala Mk Is and Mk IIs, 130 Harvards, 34 Bosboks, 20+ Kudus, 20 Cessna 185s, 19 Albatrosses, four Boeing 707-320s, seven Hercules C-130Bs, nine Transall C-160s, four DC-4 Skymasters, 39 C-47/DC-3 Dakotas, four Hawker Siddeley HS-125-400Bs, two Falcon 50s, eight Wasp helicopters, 95 Puma/Oryx helicopters, 14 Super Frelon helicopters and 70 Alouette III helicopters.

By 1994, with the major change in political power in South Africa, that strength had dropped to 636 fixed-wing aircraft and helicopters of 26 types. Since the end of border hostilities in 1989, several squadrons had been disbanded, a number of bases and units had been closed down, and all the remaining Buccaneers, Canberras, Transall C-160s, DC-4 Skymasters, Bosboks, Kudus and Albatrosses, in addition to the remaining Super Frelon and Wasp helicopters, had been withdrawn from service.

From November 1994 onwards, the SAAF started operating the first of 60 Swiss-manufactured Pilatus PC-7 Mk II Astra basic trainers, replacing the stalwart Harvards. In 1998, the SAAF received 12 locally produced Denel 'Rooivalk' (Afrikaans for Red Hawk) attack helicopters.

The Mirage F1s were withdrawn from service in November 1997 when SAAF No. 1 Squadron was disbanded, and in 2004 a British Aerospace Hawk Mk 120 Lead-In Fighter Trainer (LIFT) aircraft was delivered to South Africa, followed by a further 23 for final assembly by Denel. The remaining two Boeing 707-320C in-flight refuelling aircraft were mothballed in November 2007, and in April the following year, the last of the Cheetahs was retired.

In 2005, the SAAF announced a seven-year acquisition programme to purchase 26 Swedish SAAB JAS 39 Gripen jets (including nine two-seater trainers), 24 BAe Hawk advanced trainers, 30 Agusta A-109 light utility helicopters and four AgustaWestland Super Lynx multi-role combat helicopters.

A brace of SAAF No. 2 Squadron Advanced Light Fighter SAAB Gripens, first introduced in 2008.

Other books you might like:

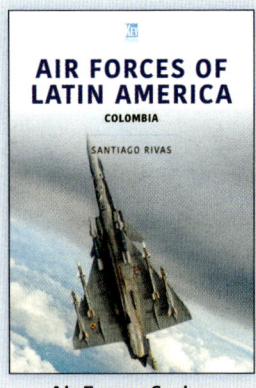
Air Forces Series, Vol. 5

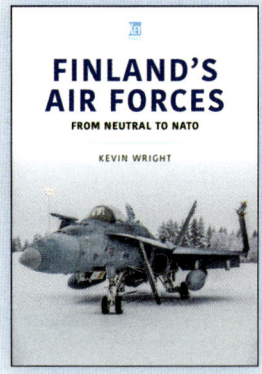
Air Forces Series, Vol. 6

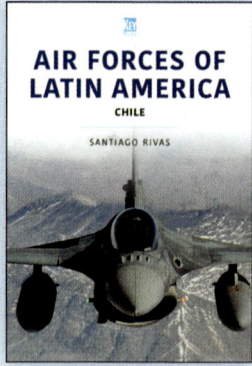
Air Forces Series, Vol. 7

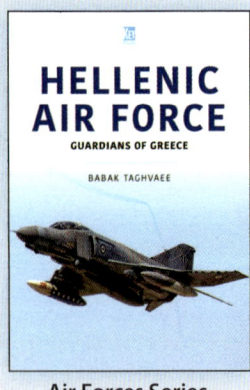
Air Forces Series, Vol. 8

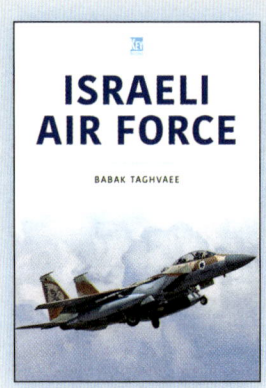
Air Forces Series, Vol. 10

Air Forces Series, Vol. 11

For our full range of titles please visit:
shop.keypublishing.com/books

VIP Book Club

Sign up today and receive
TWO FREE E-BOOKS

Be the first to find out about our forthcoming book releases and receive exclusive offers.

Register now at **keypublishing.com/vip-book-club**

Our VIP Book Club is a 100% spam-free zone, and we will never share your email with anyone else. You can read our full privacy policy at: privacy.keypublishing.com